CRYOPRESERVATION IN EUKARYOTES

Edited by **Francisco Marco-Jiménez**
and **Hülya Akdemir**

Cryopreservation in Eukaryotes
http://dx.doi.org/10.5772/62605
Edited by Francisco Marco-Jiménez and Hülya Akdemir

Contributors

Yusuf Bozkurt, İlker Yavaş, Michal Jeseta, Jana Zakova, Pavel Ventruba, Igor Crha, Bartosz Kempisty, Stefano Cinotti, Maura Ferrari, Tina Lombardo, Sabrina Renzi, Silvia Dotti, Raycon Garcia, Danilo Pedro Streit Junior, Leandro Cesar Godoy, Elsa Cabrita, Alfredo Ramírez-Reveco, Jorge Luis Hernández, Pablo Aros, Mitchel Schiewe, Nurhusien Yimer, Asmatullah --- Kaka, Rosnina Yusoff, Abdul Wahid Haron, Avelina Sotres-Vega, Jaime Villalba-Caloca, Miguel O. Gaxiola-Gaxiola, Miriam Mendoza-Santiago, Juan A. Morales-Tiburcio, Claudia Hernández-Jiménez, Francisco Figueroa-Cavero, Grace Murilla, John Thuita, Rung Yi Wu, Shao-Yu Chang, Ting-Fang Hsieh, Yen-Hsu Lai, Yu-Sen Chang, Keng-Chang Chuang, Ie Ting, Melina Bertol, Yakov Bentov, Julia Szeptycki

Published by InTech

Janeza Trdine 9, 51000 Rijeka, Croatia

Publishing Process Manager Maja Bozicevic

Technical Editor SPi Global

Cover InTech Design team

First published December, 2016
Printed in Croatia

Additional hard copies can be obtained from orders@intechopen.com

Cryopreservation in Eukaryotes, Edited by Francisco Marco-Jiménez and Hülya Akdemir

p. cm.

Print ISBN 978-953-51-2779-6
Online ISBN 978-953-51-2780-2

For EU product safety concerns:
IN TECH d.o.o., Prolaz Marije Krucifikse Kozulić 3, 51000 Rijeka, Croatia,
info@intechopen.com or visit our website at intechopen.com.

Meet the editors

Professor Francisco Marco-Jiménez holds an Agronomy Engineer degree (2003) and a PhD degree (2007) from the Polytechnic University of Valencia (Spain), where he is currently working as Professor of Animal Physiology and Biotechnology of Reproduction in the Institute for Animal Science and Technology. His research interests are mainly related in reproductive biology and technology specifically in embryo and gamete cryopreservation.

Dr. Hülya Akdemir holds a BS degree in Biology from the Istanbul University (Turkey) and MSc and PhD degrees from Gebze Technical University (Turkey), where she is currently working as a researcher. Her research interests are mainly related to different aspects of plant biotechnology including micropropagation and cryopreservation of ornamental and woody plants, assessment of genetic integrity in the plants by DNA-based molecular markers, development of genetic transformation techniques, as well as gene expression profiling of cell wall–related genes.

Contents

Preface

Since accidentally discovering the ability of glycerol on protecting cells from freezing damage, many researchers have been pursuing to develop cryopreservation methods of a very wide range of cells and some tissues. Cryopreservation is a useful tool for long-term maintenance of genetic resources of several organisms together with its contribution on conservation of their biodiversity because these methods allow protection of cells, tissues, and organs at ultralow temperatures (usually at –196 °C in liquid nitrogen) for unlimited periods. Cryopreservation especially had a great impact on reproductive medicine and recently on plant science, and it led to the establishment of cryobanks in several countries in the world with the development of well-optimized and widely used protocols. Even cryopreservation techniques represent the safest alternative for long-term preservation, because there is no "universal protocol" for all cells and tissues; conservation studies are still limited with certain species.

Cryopreservation in Eukaryotes demonstrates many different ways in which advances in cryopreservation have merged for cells and some tissues, and it totally includes 12 chapters which have been written by the expert researches in the field. All the chapters are a comprehensive collection of the most frequently used cryopreservation techniques in eukaryotes. The book chapters fall into mainly five sections. Section I ("Parasite Cryopreservation") including one chapter describes the trypanosoma preservation. Section II ("Cell and Tissue Cryopreservation") with two chapters describes recent developments to cryopreserve mesenchymal stromal cells and cartilage. Section III ("Sperm Cryopreservation") with five chapters describes different approaches on cryopreservation of fish and mammal sperms. Section IV ("Oocyte and Embryo Cryopreservation") including three chapters describes an overview of the current state. Finally, Section V ("Plant Cryopreservation") including one chapter describes recent developments in cryopreservation of orchid germplasm. Some chapters give basic principles of cryopreservation and popular techniques used, while the others combine the literature reviews and case studies.

With this book, every researcher will better understand the principles, background, and current status of cryopreservation in particular organisms. We would like to thank all the authors for their distinguished contributions, InTech Publishing Company, and its Publishing Process Manager Ms. Maja Bozicevic for her great patience during the preparation of this book.

Prof. Francisco Marco-Jiménez
Universitat Politècnica de Valencia
Institute for Animal Science and Technology
Valencia, Spain

Dr. Hülya Akdemir
Gebze Technical University
Department of Molecular Biology and Genetics
Kocaeli, Turkey

References

[1] Gonzalez-Arnao MT, Martinez-Mantero ME, Cruz-Cruz CA, Engelmann F. 2014. Advances in cryogenic techniques for the long-term preservation of plant biodiversity. In: Biotechnology and Biodiversity, Sustainable Development and Biodiversity 4 (Eds. MR Ahuja, KG Ramawat), Springer International Publishing, Switzerland, pp. 129-170.
[2] Pegg DE. 2002. The History and Principles of Cryopreservation. Seminars in Reproductive Medicine, 20(1):5-13.
[3] Reed BM. 2008. Plant Cryopreservation: A Practical Guide, Reed BM (Ed.) Springer Science and Business Media LLC, New York.
[4] Sarasan V, Cripps R, Ramsay MM, Atherton C, McMichen M, Prendergast G, Rowntree JK. 2006. Conservation in vitro of threatened plants-Progress in the past decade. In Vitro Cell Dev Biol Plant, 42:206–214.

Section 1

Parasite Cryopreservation

Chapter 1

Isolation and Cryopreservation of Trypanosomes and their Vectors for Research and Development in Resource-Constrained Settings

Murilla Grace, Ndung'u Kariuki, Joanna Auma,
Purity Gitonga and Thuita John

Additional information is available at the end of the chapter

http://dx.doi.org/10.5772/65283

Abstract

Biorepositories for biological samples have increasingly become very important in supporting biomedical research since the 1990s. The Kenya Trypanosomiasis Research Institute Cryo-bank for trypanosomes and their vectors was established in the 1970s with the aim of providing research materials to scientists. Over 2000 trypanosome isolates have been collected and stored in dewars under liquid nitrogen. Recent collections include tsetse flies—vectors of human and animal trypanosomiasis. Challenges encountered include distances to remote field sites and impassable roads and the cost of collection, preparation, storage, and maintenance under resource-constrained settings. Under these settings, the challenges can be overcome through strategic leadership that ensures availability and sustainability of resources, appropriate institutional policies, adoption of multidisciplinary approach where appropriate, working with different sectors such as human health, livestock, and wildlife, and environmental conservation in order to leverage on capacities in these sectors, and acknowledging the role of communities from which materials are collected.

Keywords: cryo-bank, cryopreservation, trypanosomes, stabilates, tsetse flies

1. Introduction

Cryopreservation is an established practice of freezing and storing valuable biological materials in liquid nitrogen for long periods of time for use in research, medicine, environmental studies, and technology development. These materials include parasites, vector tissues, and organs, a wide range of human stem cells, plants, microorganisms, etc. Research on these materials assists in understanding how ecosystems function, how disease transmission takes place, how human bodies function, and why some vectors of the same species are efficient at disease transmission whereas others are not able to transmit. Recognizing the importance of collections of biological materials to research and development and acknowledging the high cost of field sample collection in terms of financial resources and time, the management of the Kenya Trypanosomiasis Research Institute (KETRI) has put in place an institutional policy of continuous collection of trypanosome parasites for cryopreservation. This took advantage of all field visits undertaken by various scientific research teams to different foci in Kenya, a country that is endemic for both human and animal trypanosomiasis. This resulted in the establishment of the KETRI Trypanosome Bank which currently has over 2000 isolates [1] from various hosts (tsetse flies, human, domestic and wild animals). Some of the recent collections include vectors of trypanosomiasis, the tsetse flies. Updating of the cryo-bank with fresh trypanosome isolates is a continuous process.

Trypanosomes are extracellular protozoan parasites which cause debilitating disease in humans and animals. In humans, the disease is referred to as human African trypanosomiasis (HAT) or sleeping sickness, caused by two trypanosome species, *Trypanosoma brucei gambiense*, responsible for the chronic form of HAT in West and Central Africa, and *T. b. rhodesiense*, which causes acute disease in eastern and southern Africa. The parasites are transmitted by tsetse flies (*Glossina* spp.). In animals, the disease is referred to as African animal trypanosomiasis (AAT; nagana in cattle, sheep, and goats; surra in camels) and is caused by various trypanosome species, the major ones being *T. vivax*, *T. congolense* and *T. evansi* [2]. Whereas majority of the trypanosome species which cause AAT are transmitted by tsetse fly vectors, *T. evansi*, is transmitted mechanically by biting flies such as *Tabanus* spp.; *T. vivax* has been reported to be transmitted by both tsetse flies and biting flies [3].

HAT is classified in the category of the most neglected tropical diseases. Current diagnostic tools have inadequate sensitivity and specificity, thus complicating disease diagnosis and staging. The drugs available for treatment are highly toxic and not very effective; patients die if untreated [4, 5]. In 2005, an annual prevalence of 50–70,000 HAT cases/year was reported, with incidence rates of 15–17,000 cases/year [6]. Although recent data from the World Health Organization (WHO) shows that the number of reported cases of HAT declined to less than 10,000 in 2009 leading to speculation that the disease could be eliminated [7, 8], there is great need to maintain vigilance through surveillance and research. This is informed by the fact that HAT was effectively controlled in the 1960s in many endemic countries; however, the disease re-surged due to breakdown in surveillance and control activities (**Figure 1**). WHO [9] has developed a roadmap for elimination of HAT by the year 2020, which involves development of new and better diagnostics and drugs [5, 8]. Cryo-banks such as the KETRI Trypanosome Bank will therefore be important in contributing to this strategy in order to ensure that epidemics do not occur in future; and that dormant foci will be prioritized for elimination. One of the issues for which answers are sought is what happens in some traditional HAT foci when the disease is not reported in humans. Some of the new technological advances that are providing more insights include genetic analysis of both parasite and vector genomes and identification of specific proteins as targets for development of vaccines, new and sensitive diagnostic tests.

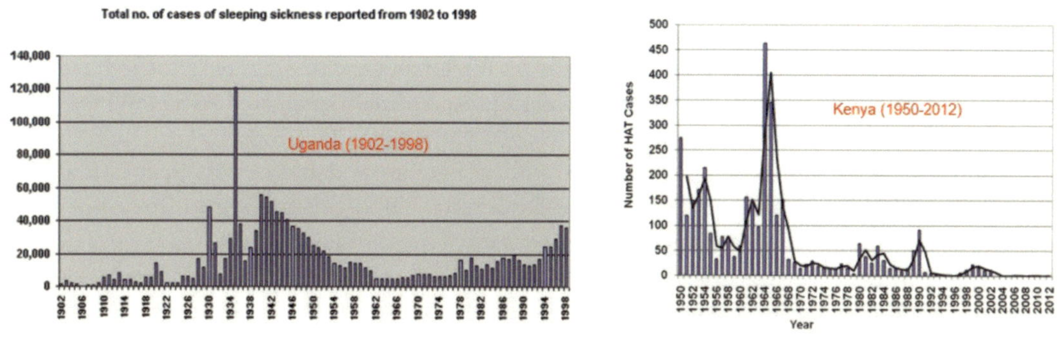

Figure 1. Sleeping sickness as a reemerging disease.

Isolation and cryopreservation of new trypanosome strains from patients in different HAT foci ensures availability of these stabilates for use in parasitological, biochemical, molecular, serological and pharmacological investigations many years after their isolation from the host. Brun *et al* [2] observed that one of the major obstacles in the elucidation of the factors responsible for relapses after melarsoprol treatment was the lack of recent *T. b. gambiense* isolates from patients from various endemic areas where the problem has been reported. The WHO steering committee on human African trypanosomiasis treatment has therefore recommended that collection of stabilates be a continuous activity in order to monitor the occurrence and spatial distribution of treatment failure [10] and refractoriness of tsetse to infection. Since its inception, KETRI and now KALRO-Biotechnology Research Institute developed an institutional policy of encouraging collection of stabilates by scientists and clinicians, for cryopreservation. In this chapter, we describe the procedures of isolation and cryopreservation of trypanosome stabilates for research and development in resource constrained settings.

2. Field isolation of trypanosomes

Trypanosomes are isolated from infected hosts during active or passive disease surveillance activities. The infected hosts include humans, domestic and wild animals, as well as tsetse fly vectors. Parasites are isolated from biological fluids including blood, cerebrospinal fluid (CSF) and lymph node aspirates, and/or body parts of tsetse fly vectors. Depending on the host parasitemia and/or density of trypanosomes in the biological fluids at the time of isolation, trypanosomes can be either cryopreserved directly or propagated in immunosuppressed laboratory rodents prior to cryopreservation.

2.1. Diagnosis: buffy coat and whole blood parasitemia

Parasitological diagnosis of trypanosome infections in animals and humans can be made through microscopic examination of wet blood smears, stained thin and thick blood smears, smears of lymph node aspirates, and buffy coats [11]. Under normal field conditions when large numbers of animals are sampled, examination of buffy coats, obtained through capillary tube centrifugation technique (CTC) [12], is the preferred method of diagnosis due to its higher sensitivity compared to other microscopic techniques. Animals suspected to be infected with trypanosomes are bled from the ear vein into heparinized capillary tubes after which the 3/4 full capillary tubes are sealed at one end with plasticine and then spun in a hematocrit centrifuge at 10,000 revolutions per minute for 5 min. Blood separates into three portions, namely, the red blood cells, which settle at the bottom of the capillary, the plasma portion found at the top, and the buffy coat portion, which forms at the interface of the red blood cells and plasma. Trypanosomes are concentrated in the buffy coat portion of the centrifuged blood, thus enhancing the sensitivity of the test. In humans, diagnostic methods that are routinely employed to detect blood trypanosomes include (CTC), quantitative buffy coat (QBC), mini anion exchange centrifugation technique (mAECT), and modified mAECT [13]. For the diagnosis of trypanosomes in cerebrospinal fluid, available methods include single and double centrifugation and modified single centrifugation (MSC), with the MSC being easy to perform and as sensitive as the double centrifugation [13].

Once confirmed positive, the density of trypanosomes in the relevant biological, fluid is determined. Whole blood is drawn from the jugular vein of the infected host into anticoagulant containing tubes and used to quantify the parasitemia using the matching method [14] for the Trypanozoon group of trypanosomes. Direct isolation is therefore determined by whole blood parasitaemia.

2.2. Direct isolation of parasites from infected biological fluids

Parasitemia is usually low in naturally infected hosts. However, the required density of between 3.2×10^7 trypanosomes/ml and 1.3×10^8 trypanosomes/ml may be obtained in a small proportion of the infected hosts, thus permitting direct cryopreservation of the stabilates. In such cases, the infected whole blood is mixed with either of the following cryoprotectants and processed:

1. 20% glycerol in EDTA saline glucose (ESG), pH 8.0 in the ratio of 1:1.
2. Glycerol in the ratio of 1:4, that is, one part of the infected blood to four parts of glycerol.

Samples are then labeled and dipped into a vapor shipper liquid nitrogen cylinder for transportation from the collection site to the main laboratory for further processing. A vapor shipper is a liquid nitrogen cylinder which has a mechanism of absorbing liquid nitrogen into its system leaving the hollow space full of liquid nitrogen vapor, with temperature in the range of –60 to –80°C. In the laboratory, samples are removed from the vapor shipper and allowed to thaw on ice or at 4°C after which they are dispensed into plain capillary tubes (**Figure 2**), which are then sealed at one end using plasticine.

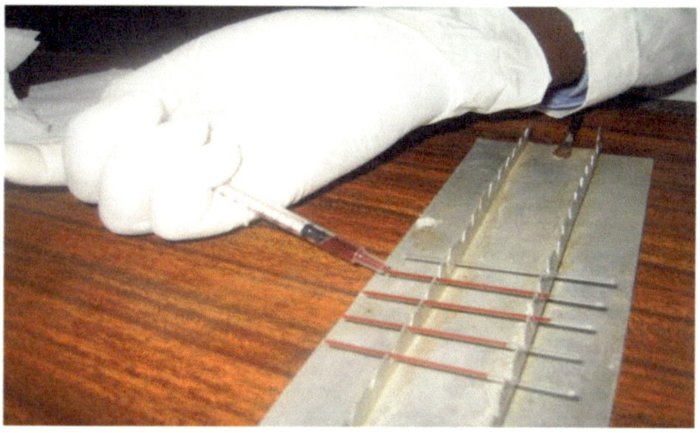

Figure 2. Loading plain capillary tubes with cryo-protected infected blood sample.

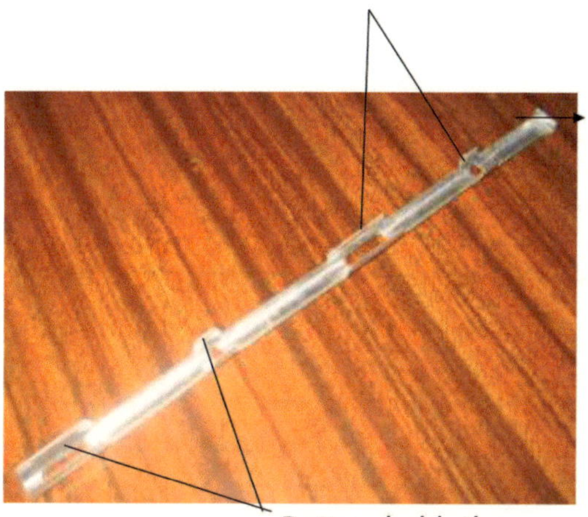

Figure 3. Aluminum cane for holding the ampoules.

The loaded and sealed capillary tubes, approximately 18 in number, are then accommodated in a perforated 4.5 ml ampoule tube into which a label is inserted. The label has the laboratory sample identification number. The ampoule normally has two perforations, one at the top and the other at the bottom to allow direct contact of the sample with liquid nitrogen and at the same time continuous flow of liquid nitrogen from the top through the bottom perforations. Before permanent storage, a capillary is removed from the suspended sample and the viability of the trypanosomes confirmed. For ease of permanent storage in liquid nitrogen at –196°C, the ampoule is then placed on to an aluminum cane which has the capacity of holding at least two ampoules: one at the top and the other at the bottom of the cane. The cane is then placed into a canister, with each canister having a capacity of holding 14 canes. The canister is then immersed into liquid nitrogen in the permanent storage dewar (**Figures 3** and **4**) at –196°C.

Figure 4. Stabilate storage dewars.

2.3. Isolation and propagation of low parasitemia blood samples in laboratory rodents

When the parasitaemia in the whole blood of the naturally infected host is low i.e. below 1.3×10^8 trypanosomes/ml, the anticoagulated infected blood sample is intraperitoneally inoculated into an immunosuppressed laboratory rodent. It is advisable to inoculate two rodents, usually mice, per positive blood sample (isolate). This is carried out in the field where the animals are given identification numbers and transported to the laboratory for monitoring. In the laboratory, the infected mice are maintained on commercial mice pellets (Unga Feeds Ltd., Kenya), provided with water *ad libitum* and monitored for development of parasitemia. At the first peak of parasitemia between 1.3×10^8 trypanosomes/ml and 2.5×10^8 trypanosomes/ml, the infected mice are euthanized and the blood harvested by cardiac puncture into tubes containing EDTA as anticoagulant. The blood is mixed gently before addition of a cryopreservative at a ratio of 1:1. The samples are suspended in liquid nitrogen vapor for at least 2 h using a cooling jacket before permanent storage in liquid nitrogen at –196°C [15, 16]. Samples with a trypanosome concentration below 6.3×10^7/ml may also be preserved by direct addition of glycerol to infected biological fluids to a final concentration of 20%, especially if the isolated trypanosome species does not infect rodents [17].

3. Cryopreservation

3.1. Cryopreservation media

The most commonly used cryopreservative is 20% glycerol in EDTA saline glucose (ESG). ESG is first prepared by dissolving 8.00 g NaCl, 0.30 g KH_2PO_4, 2 g EDTA (disodium or dipotassium), and 2 g glucose in 800 ml distilled water, adjusting the pH to 8.0, and then topping up to 1 liter with deionized/distilled water. Glycerol is then added to the prepared solution at a ratio of 1:4 to make 20% glycerol in ESG. Recently, Triladyl® (MiniTüB GmbH and CO. KG, Tiefenbach, Germany), a commercially available culture medium that is traditionally used for preserving bull semen, has been found suitable for cryopreservation of trypanosomes. Triladyl is most efficient when added to the infected biological fluids/materials

http://dx.doi.org/10.5772/65283

to a concentration of 40–90% [18, 19]. Triladyl has subsequently been adopted as an alternative cryopreservative at the KETRI Trypanosome Bank.

Buffers commonly used in cryopreservation

EDTA saline glucose (ESG) pH 8.0

NaCl 8.00 g

KH_2PO_4 0.30 g

EDTA (disodium or dipotassium) 2 g

Glucose 2 g

Dissolve in about 800 ml distilled water, top up to 1 l.

Normal saline

Dissolve 8.5 g NaCl in 1 l of distilled water

Phosphate Saline glucose (PSG) pH 8.0

Na_2PHO_4 5.392 g/l

Na_2HPO_4 $(H_2O)_{12}$ (hydrous) 13.608 g/l

NaH_2PO_4 0.239 g/l

$NaH_2PO_4(H_2O)$ hydrous 0.276 g/l

$NaH_2PO_4(H_2O)_2$ hydrous 0.312 g/l

NaCl 1.7 g/l

Glucose 10.00 g

Dissolve in about 800 ml distilled water; top up to 1 l.

3.2. Cryopreservation procedure

Biological materials from infected vertebrate hosts (blood, cerebrospinal fluid (CSF), and lymph node aspirates) and/or body parts of tsetse fly vectors are processed as previously described [17, 20, 21]. Briefly, samples with a concentration of more than 6.3×10^7 trypanosomes/ml are mixed with a cryopreservative (20% glycerol in ESG or 40–90% Triladyl) at a ratio of 1:1. The diluted sample is then loaded into an ampoule and wrapped or held in a cooling jacket which is suspended into the vapor space of a liquid nitrogen shipper (temperature range of –60 to –80°C) for transportation to the laboratory. Samples from infected humans or animals with parasite counts below 6.3×10^7 trypanosomes/ml (equivalent to antilog 7.8) [14], or suspects with a low packed cell volume (usually <25%) are inoculated into laboratory rodents for amplification/propagation [19]. The rodents may be immunosuppressed using either cyclophosphamide at 100 mg/kg daily for three consecutive days or by gamma-irradiation at 600 rads (6 gray) for 6 min before the inoculation [22, 15]. Swiss white mice are preferred for propagation of most species of trypanosomes, while Mastomys natalensis are preferred for *T. b. gambiense* [23]. Inoculated rodents are sacrificed at the first peak of parasitemia and the blood harvested by cardiac puncture into tubes with EDTA as anticoagulant. The blood is mixed gently before addition of a cryopreservative at a ratio of 1:1. The diluted samples are then suspended in liquid nitrogen vapor for at least 2 h before being stored permanently in liquid nitrogen at –196°C [24, 25].

Essential requirements for cryopreservation

1. Immunosuppressed mice
2. Mice pellets
3. Sawdust
4. Disposable 1 ml syringes and needles gauge 26
5. Trypanosomes either from the cryo-bank or from infected hosts (humans/animals) or tsetse flies
6. Glycerol
7. Cotton wool
8. Tissue paper
9. Capillary tubes (plain)
10. Plasticine
11. 4.5 ml plastic ampoules
12. Liquid nitrogen
13. Cooling jacket

The following information is collected and recorded:

1. Host of isolation
2. Locality (georeferenced) and year of isolation
3. Scientist who did the isolation
4. Number of passages the trypanosomes has undergone before cryopreservation

5. The prepatent period of the infected mice after each passage, if several passages have been done
6. The duration of infection in the infected mice between the time of inoculation and sacrifice of the mouse for cryopreservation
7. Species of trypanosome
8. Physical location of the sample in the cryo-bank

In event of cryopreservation from the natural host:

- The suspected infected blood sample is injected into immunosuppressed mice
- The infected mice are monitored for development of parasitemia
- Cryopreservation carried out as outlined above

At the KETRI trypanosome cryo-bank, the first parasite population isolated from the field is the original or primary isolate. If the parasite numbers are high, the primary isolate is cryopreserved as original field isolate; however, in order to sustain/maintain the original cryopreserved sample and to produce adequate material for research, the original isolate is expanded in the appropriate animal model and cryopreserved as a derivative of the original sample, but with a different bank reference number from the first passage number. Subsequent passages or derivatives of the same are given different reference numbers in order to monitor the use and performance of the particular isolates. Clones may be prepared either from the primary or subsequent passages or both. All these events are monitored and recorded for future reference and when issuing materials for research.

3.3. Preparation of the cryopreserved sample for the infection of laboratory animals

Following receipt of the duly signed and approved request form from the scientist, the person incharge of issuance of cryopreserved material records the physical position of the sample in the cryo-bank. The sample is retrieved from its position, the ampoule cork opened; and using a pair of forceps, the reference number is confirmed. One capillary is removed and transferred into an ampoule placed on ice for thawing. The remaining stabilate is returned into its position in the bank before it thaws.

The procedure for infection of laboratory animals

1. Use 1 ml syringe and 26-gauge needle
2. Fill the syringe with phosphate saline glucose (PSG) pH 8.0 up to 0.2 ml
3. Using a diamond pencil cut the sealed end of the capillary tube containing the trypanosome stabilate
4. Insert the needle into the capillary and suck the contents

5. Pull the piston and mix the contents with the PSG buffer thoroughly

6. Remove the air bubbles and place a drop of the mixture on the microscope slide, cover with a cover slip, and examine the parasitemia at 400 × magnification

7. Infect the experimental (donors) as required by the protocol

Precaution:

1. Use of protective devices while handling cryopreserved samples is mandatory. This is necessary because the samples being handled contain live parasites, some of which are pathogenic to humans. Also, nitrogen at –196°C burns. Industrial gloves are recommended while handling liquid nitrogen. Use of facial masks will protect the user from harmful effect in case of contact with the eyes. All safety precautions should be strictly observed when capillary tubes are withdrawn from the liquid nitrogen; they sometimes burst before they are transferred into the screw capped ampoules to thaw, possibly due to differences in temperature.

2. It is recommended that retrieval of the sample should be rapid to avoid the thawing of the remaining samples.

4. Cloning of trypanosomes

Cloning of trypanosomes is necessary for the production of a homogeneous population of trypanosomes. It is carried out as previously described by Otieno and Darji [23]. Briefly, a sample containing trypanosomes is diluted in PSG pH 8.0, to 1 trypanosome per microscopic field at 400× magnification. This is followed by addition of 0.5 ml guinea pig serum. Using a needle, a drop of the trypanosome suspension is placed on a cover slip that is overturned onto a cavity slide moistened with PSG to prevent evaporation of the drop. In the laboratory, the drop is examined under a microscope (400× magnification) by at least three experienced technicians to confirm the presence of a single and viable trypanosome, which is aspirated and inoculated into an immunosuppressed Swiss white mouse. Inoculated mice are monitored for parasitemia development and sacrificed at the first peak of parasitemia. Blood is harvested by cardiac puncture for cryopreservation of the clone of trypanosomes.

5. Maintenance of cryopreserved trypanosomes

5.1. Liquid nitrogen

Cryopreserved trypanosomes are permanently stored in liquid nitrogen. The samples must always be fully immersed in liquid nitrogen and the levels maintained by frequently refilling the storage Dewars. The refilling period is determined by the frequency at which the Dewars are opened during issue of materials for research. The more frequent they are opened, the more liquid nitrogen vapor is lost, hence the need to refill. Under normal circumstances, refilling is done fortnightly.

5.2. Trypanosomes viability and infectivity tests

It is important to ascertain that the cryopreserved samples remain viable by randomly testing the infectivity of the parasites in laboratory rodents to ensure that this has been maintained and not lost over long periods of storage. The viability testing is performed by removing a single capillary of each of the cryopreserved trypanosome isolate, thawing at room temperature, cutting the sealed end of the capillary tube using a diamond pencil, decanting the capillary contents on a microscope slide, covering the content with a glass coverslips, and examining for the motility of the trypanosomes under the microscope at 400× magnification.

6. Morphological characterization of trypanosomes

Morphological features assist in the preliminary identification of trypanosomes in the field after isolation in order to ascertain the species of trypanosomes isolated. This is done through microscopic examination. Different trypanosome species fall into the following groups, depending on the morphological features: Trypanozoon, Duttonella, or Nannomonas (**Table 1**).

Species	Morphology	Free flagellum	Undulating membrane	Kinetoplast	Other characteristics
T. vivax	Monomorphic	Present	Slightly developed	Large rounded terminal	Very motile, posterior end rounded
T. uniforme	Monomorphic	Present	Slightly developed	Large rounded terminal	Very motile, posterior end rounded
T. congolense	Pleomorphic	Absent	Slightly developed	Marginal or central subterminal	Posterior end rounded or flat
T. simae	Pleomorphic	Absent	Moderately developed	Marginal or central subterminal	More long forms than short forms
T. brucei	Pleomorphic	Present in long and intermediate forms, absent in short forms	Well developed	Small subterminal	Posterior end: Long forms—pointed Short forms—rounded Intermediate forms—blunt
T. evansi	Basically monomorphic	Present	Well developed	Small subterminal	Posterior end rounded or truncated
T. suis	Monomorphic	Present	Well developed	Small, subterminal, marginal	Only infects suids
T. theileri	Monomorphic	Present	Well developed	Large, rounded, marginal	Posterior end tapering

Table 1. Morphological characteristics of trypanosomes.

6.1. Some of the trypanosome species distinguishing morphological characteristics include:

- Size and shape of the body
- Position of the nucleus and kinetoplast
- Presence or absence of free flagellum
- The shape of the posterior end which is pointed either sharply, oval, or blunt

The morphological features of the Trypanozoon, Duttonella, and Nannomonas species of trypanosomes are as shown in **Figures 5–7**.

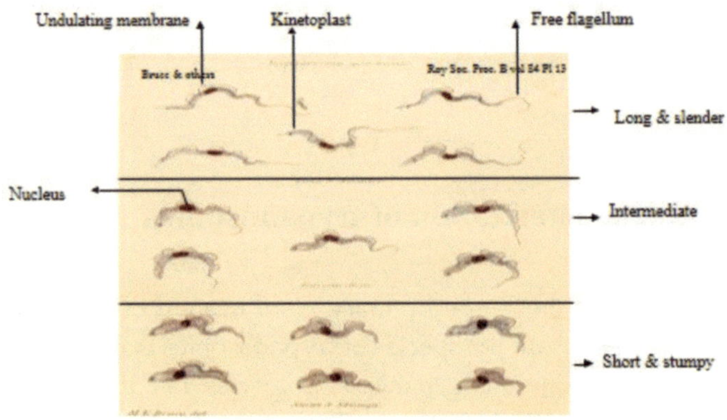

Figure 5. Morphological distinguishing features associated with Trypanozoon.

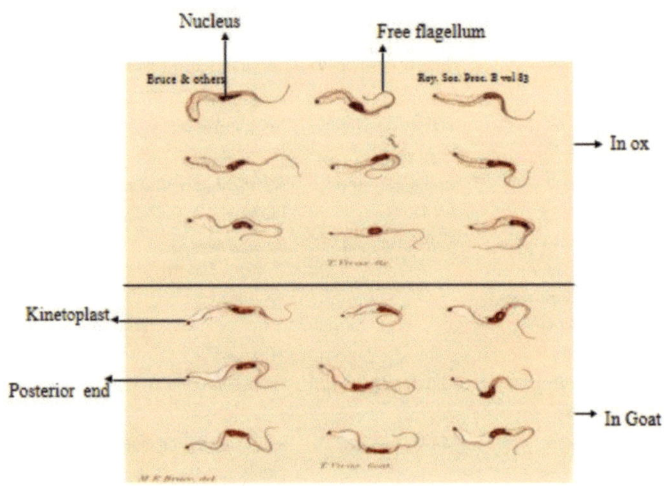

Figure 6. Morphological distinguishing features associated with Duttonella.

http://dx.doi.org/10.5772/65283

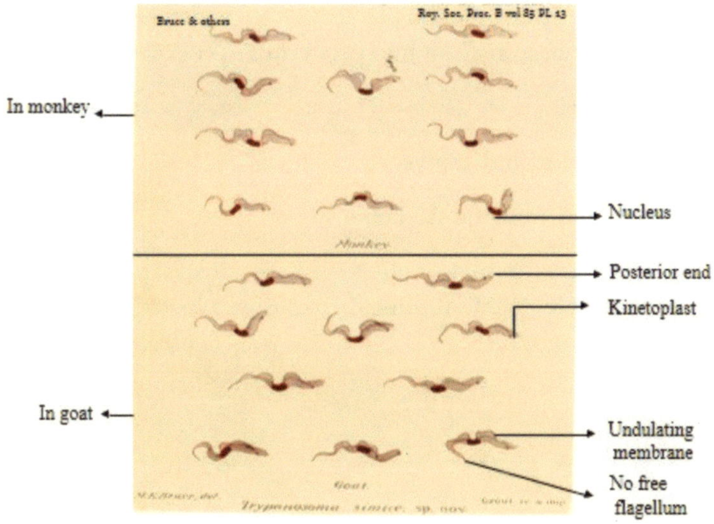

Figure 7. Morphological distinguishing features associated with Nannomonas.

6.1.1. Trypanosoma brucei gambiense (Trypanozoon) subgroup

Morphological distinguishing features associated with Trypanozoon include:

- Position of kinetoplast—subterminal
- Size of kinetoplast—small
- Posterior end—blunt
- Free flagellum—present

6.1.2. Trypanosoma vivax (Duttonella) subgroup

Distinguishing features include: rounded posterior end, free flagellum, and a large almost terminal kinetoplast. This group of parasites lack undulating membrane.

6.1.3. Trypanosoma simiae in the same group with T. congolense (Nannomonas)

Distinguishing features include: shape of posterior end—rounded; position of kinetoplast — marginal; size of kinetoplast—medium; no undulating membrane and free flagellum—present.

6.1.4. Other methods employed in the laboratory to characterize trypanosomes

- Tsetse transmission studies

  Trypanosoma evansi is not transmitted by *Glossina* but by biting flies.

- Tissue culture

  Trypanosoma evansi does not transform into procyclics in culture.

- Molecular biology studies:
  - DNA and RNA extraction and analysis
  - Genetics and functional genomic studies
- *In vivo* studies using various animals models to characterize virulent phenotypes and drug resistance. It is important to note that some trypanosome species such as *Trypanosoma vivax* do not grow in rodents.

7. Documentation

Over the years of existence of the KETRI Cryo-bank, the cryopreserved materials were documented manually on specially designed record sheets known as Kalamazoo and later electronically. Whichever method is used, the information in the records should include, but not limited to the following, for the ease of retrieval:

- The host of isolation, age and sex
- Date of isolation
- Isolating scientist
- Suspected species of trypanosome (based on host of isolation and trypanosome morphology)
- Locality of isolation
- Method of cryopreservation whether direct or after propagation in laboratory animals
- If by propagation in laboratory rodents, the species of rodent used
- The prepatent period, i.e., period between inoculation of the whole blood and first appearance of trypanosomes in the mouse
- Duration of infection before cryopreservation, i.e., the period between inoculation and the sacrifice of the animal to harvest trypanosomes for cryopreservation
- In the event where several passages have been made, the passage numbers must be indicated as well as the species of rodents used at every passage. In addition, the pre-patent period and duration of infection must be stated at each passage.
- Physical location of the stabilate in the cryo-bank
- Work carried out and publications resulting from the use of stabilates

http://dx.doi.org/10.5772/65283

7.1. Electronic database

The database was developed using Microsoft Access 2000 (Microsoft, USA) relational database. Hosts of isolation and countries are coded following the International Organization for Standardization (ISO) protocol [26] (Lumsden, 1978). Primary isolates are, for example, designated MHOM/KE/85/KETRI 128, where M represents mammal; HOM represents human; KE indicates the country of isolation, in this case Kenya; 85 is the year of isolation (1985) while KETRI 128 shows code or reference number of the stabilate. With regard to trypanosome derived from the previous stabilate usually referred to as derivatives, the number of the derivative is shown in brackets. For example, MHOM/KE/85/KETRI 128 [KETRI 300] is a derivative of MHOM/KE/85/KETRI 128 as described by Lumsden [25].

7.2. Trypanosomes isolated, documented, and cryopreserved at KETRI Cryo-bank

The number of trypanosomes isolated and cryopreserved at the KETRI Trypanosome Bank are shown in **Figure 8**. **Tables 2** and **3** show the trypanosome species, year, and country of isolation.

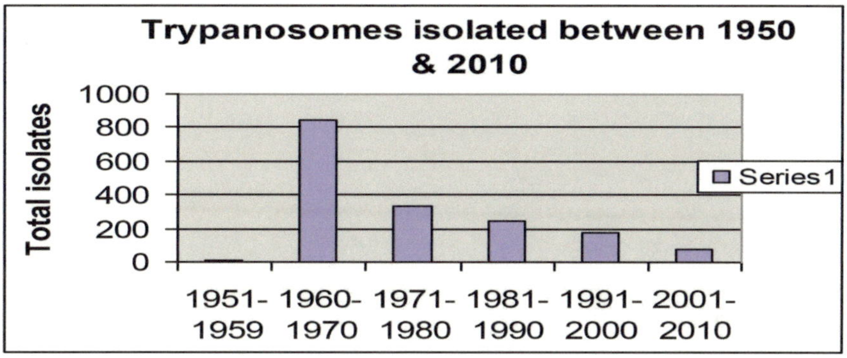

Figure 8. Number of primary trypanosome stabilates collected, preserved, and stored at Kenya [1].

		Species of trypanosomes: number and period of isolation											
Country	Isolate/ Year	*Tbb*	*Tb subgroup*	*Tbr*	*Tbg*	*T. congolense*	*T. vivax*	*T. evansi*	*T. simiae*	*T. theileri*	*T. lewesi*	UC	Mixed
Kenya	No	101	194	274	–	107	166	89	3	–	–	–	29
	Year	1961–2001	1961–2006	1958–2009	–	1961–2008	1969–2009	1968–2003	1970	–	–	18	1970–2006
Uganda	No	1	238	123	22	82	64	–	–	2	8	14	5
	Year	1968	1960–1983	1959–2004	1959–2002	1955–1983	1961–1972	–	–	1972–1973	1966	14	1955–1971
Tanzania	No	–	57	7	–	35	–	–	–	–	–	–	9
	Year	–	1966–1974	1934, 1959–1974	–	1966–1974	–	–	–	–	–	–	1966–1974
Botswana	No	–	–	2	–	–	–	–	–	–	–	–	–
	Year	–	–	1960		–	–	–	–	–	–		–
Sudan	No	–	–	–	26	–	–	2	–	–	–	–	–
	Year	–	–	–	1982–2003	–	–	1973	–	–	–	–	–

		Species of trypanosomes: number and period of isolation											
Country	**Isolate/ Year**	***Tbb***	***Tb subgroup***	***Tbr***	***Tbg***	***T. congolense***	***T. vivax***	***T. evansi***	***T. simiae***	***T. theileri***	***T. lewesi***	**UC**	**Mixed**
Mozambique	No	–	–	2	–	–	–	–	–	–	–	–	–
	Year	–	–	1980, 1983	–	–	–	–	–	–	–	–	–
Nigeria	No	–	–	–	–	–	4	–	–	–	–	–	–
	Year	–	–	–	–	–	1970–1973	–	–	–	–	–	–
Zambia	No	–	–	–	–	2	–	–	–	–	–	–	–
	Year	–	–	–	–	1981	–	–	–	–	–	–	–
NDA	No	2	21	8	–	17	5	1	–	–	–	3	2
	Year	1961	–	–	–	1962–1985	1961	–	–	–	–	–	2
Total		104	510	416	48	243	240	92	3	2	8	35	45

Adapted from Murilla et al. [1].

Tbb, *Trypanosoma bruceibrucei*; *Tbr*, *Trypanosoma bruceirhodesiense*; *Tbg*, *Trypanosoma bruceigambiense*; UN, unclassified; NDA, no data available.

Table 2. Primary trypanosome isolates collected from various countries and stored at the Kenya Trypanosomiasis Research Institute Cryo-bank

	Tbb	*Tb subgroup*	*Tbr*	*T. congolense*	*T. vivax*	*T. evansi*	*T. theileri*	*T. simiae*	*T. lewesi*	*UC*	*Mixed*	*Total*
Cattle	85	247	0	119	137	0	2		0	10	21	621
Goat	0	6	0	6	6	0	0	0	0	0	1	19
Sheep	1	8	1	24	3	0	0	0	0	0	7	44
Pig	1	5	0	0	0	0	0	0	0	0	0	6
Camel	0	3	0	1	2	92	0	0	0	0	0	98
Donkey	0	1	0	1	0	0	0	0	0	3	0	5
Cat	0	0	0	0	0	0	0	0	0	0	2	2
Dog	3	6	0	1	0	0	0	0	0	1	0	11
Wildlife	2	40	3	12	0	0	0	0	0	0	5	62
Lizard	0	0	0	0	0	0	0	0	0	3	0	3
Rat	0	0	0	0	0	0	0	0	8	0	0	8
HNI	1	7	4	2	2	0	0	0	0	2	0	18
Total	93	323	8	166	150	92	2	0	8	19	36	897

Adapted from Murilla et al. [1].

Table 3. Animal hosts from which various trypanosomes were isolated and stored at Kenya Trypanosomiasis Research Institute Cryo-bank

8. Issuance of stabilates

Trypanosomes are stored and maintained permanently in liquid nitrogen for use by scientists in the following areas:

1. The pathogenicity/virulence studies
2. Molecular characterization

3. Drug sensitivity studies
4. Tsetse fly transmission/vector competence studies, etc.
5. Genetic and genomic studies
6. Research and development of new diagnostic tests, drugs, and vaccines

Stabilates are issued to National and International Research Organizations and Institutions of higher learning following the laid down institutional guidelines. The following documents are necessary for issue of materials to be effected:

Issuance of trypanosome stabilates

Procedure

1. Research permit from National Council for Science, Technology and Innovation (NACOSTI, Kenya)
2. Stabilate requisition form plus a permit from Director of Veterinary Services (DVS) for stabilates to be used in other National and International institutions
3. Materials transfer agreement duly signed
4. Stabilate requisition form with relevant approvals for stabilates to be used only within the institute.

It is important that the receiving scientist be committed to avail the scientific information resulting from the use of these stabilates. This is necessary for updating the trypanosome bank database and for future reference.

9. Terminologies commonly associated with cryopreservation (from: Trypanosomiasis a veterinary perspective Lorne E. Stephen pages 440–441)

Terminologies commonly used with cryopreservation technique

1. Trypanosome species

   Assemblages of organisms that can be distinguished from other species by one or more stable discontinuous morphological characters, e.g., *T. congolense, T. vivax, T. brucei* are different species.

2. Trypanosome subspecies

   Assemblages of organisms within a species that cannot be separated from each other by morphological characters but only by other stable characters, e.g., *T. b. rhodesiense, T. b. gambiense* are subspecies of *T. brucei*.

3. Clone

This is a population that has been developed from a single trypanosome.

4. Line

A laboratory derivative of a stock maintained in different physical conditions, e.g., a species of *T. congolense* maintained in mice only is a line of that nature, while when maintained in cattle only is another line of that nature, etc.

5. Population

The group of trypanosomes present at a given time in a given host or it may consist of a mixture of several species and subspecies.

6. Primary isolate

This is a stabilate made from a naturally infected host or viable organisms present in a culture or experimental animals following the introduction of the sample from a naturally infected host.

7. Sample

That part of trypanosome population collected on a unique occasion.

8. Stabilate

A cryopreserved sample of viable trypanosomes.

9. Stock

A population derived by serial passage *in vivo* or *in vitro* from a primary isolate.

10. Metacyclics

These are the mammalian infective forms of trypanosomes injected by the tsetse fly. Metacyclic forms are found at the end of transmission cycle.

11. Procyclics

These are the midgut forms of trypanosomes which are found in cultures.

---

10. Conclusion

Communicable and noncommunicable diseases, including the neglected tropical diseases, cause chronic life-long disability, hinder economic development, and impair childhood development in resource poor settings in Africa where the diseases are endemic [27, 28]. Control of these diseases could be an efficient way to fight poverty since some of these diseases can be managed very cost-effectively using evidence-based control strategies [26, 29, 30]. HAT is classified as one of the most neglected tropical diseases that exclusively affects poor communities in low- and middle-income countries (LMIC), except those areas where tourists have been reported to have contracted the disease on tour of the affected areas. Because NTDs affect mostly the socially vulnerable populations, there are several ethical implications to consider when planning collection and use of these materials. Detection and treatment of these diseases poses many challenges since most of them present similar clinical symptoms concomitant with variation in response in the affected individual to treatment, and lack of accurate diagnostic tests. Medical research to improve health care faces a major problem in the relatively limited availability of adequately annotated and collected bio-specimens, primarily due to absence of bio-banking facilities and associated infrastructure to interrogate the specimen to tease out relevant information. This limitation has adversely impacted the pace of scientific advances and successful exploitation of bio-specimens. Established functional bio-banks would surmount this limitation by providing framework for transfer of bio-specimens (tissues, blood, and body fluids) and related health data for research. The KETRI Cryo-bank holds significant quantities of samples dating from 1930s to date, which include blood, serum, CSF, tissues, semen from trypanotolerant animals, and both parasite and vector DNA collections. This is in addition to the pan African trypanosome isolates of specific biomedical interest (e.g., drug resistance and virulence) from human and nonhuman primates, and livestock. There is therefore great need to collect and store biological materials for research in order to monitor our ecosystems including new and emerging diseases, generate evidence to inform policy, and in the development of mitigation strategies. In the area of human and veterinary medicine, these new and reemerging diseases and conditions have complicated the search for new remedies for their management in the absence of well collected and cryopreserved biological specimens.

10.1. The challenges

The countries that are heavy burdened by disease also experience high levels of poverty. This situation is compounded by new and reemerging diseases and conditions. Climate change has not only resulted in loss of biodiversity but enabled vectors to infest new areas and change transmission dynamics. Some parasites have changed host seeking behavior with time, becoming either more virulent or chronic in nature. Development of drug resistance and appearance of virulent phenotypes is of great public health concern. Whereas the need exists to collect and preserve these materials for R&D in order to find solutions to these challenges, the cost of sample collection from the field is prohibitive. Sites are usually remote with unpassable roads especially in rainy season when disease transmission is high. Once the materials have been collected and transported to the laboratory, there are high costs related to processing, cryopreservation, and maintenance of the cryo-bank. These are not the usual areas for investment by our governments due to different priorities. There are also challenges related to communities from which samples are collected, they are usually not involved in the plans to collect the materials thereby excluding them in finding solutions to their problems.

10.2. The opportunities

The above challenges have created great opportunities for collection and storage of parasites and their vectors for use in the development of vaccines, diagnostic tests, and new medicines applying recent technological advances. In the recent past, improvements have been made on the conventional nitrogen freezers through development and adoption of validated methods including a wide range of stem cells. Many cryopreservation protocols exist for freezing and storing various biological materials. These need to be reviewed and tailored toward delivering quality biological materials to our research institutions and products to our clinics. Modalities for sharing of materials by different institutions need to be developed and made operational in order not to disadvantage communities from which the materials are collected and the institutions that have collected, preserved, and maintained these materials in resource-constrained settings. The contractual arrangements surrounding areas of the material transfer agreements should be carefully negotiated. National and international institutions (local and foreign), should invest in adequate bio-specimen management, legal and administrative skills, just as they do for developing scientific skills to facilitate sharing of samples and information associated with the bio-specimens.

10.3. Lessons learnt: how to establish and sustain cryobank in a resource-constrained setting

Collection and cryopreservation of biological materials is critical to research and development but expensive to collect, process, store, and maintain. Institutional top management leadership supported by the existing national, regional and international guidelines, rules and regulations are necessary in providing policy direction and resources [31]. In Kenya, tsetse flies, vectors of human and animal trypanosomiasis, infest mainly conservation areas and wildlife are carriers of pathogens, hence there is a need to work closely with the Kenya Wildlife Service. Through effective collaborations and multidisciplinary approach, it is possible to leverage on all activities undertaken by collaborating institutions to make collections. From the resource-constrained perspective, one does not need state-of-the-art bio-repository to initiate collections. Strategic leadership is the key in spearheading:

1. the development of the appropriate institutional policies
    1. to define roles and responsibilities for collaborative arrangements among institutions, strictly observing existing rules and regulations
    2. to ensure that communities from which the materials are collected are not taken advantage of
    3. to facilitate the establishment of relevant multidisciplinary teams that cut across several sectors, e.g., human health, livestock, and wildlife
    4. to ensure proper collection, storage, and maintenance of the materials according to the legal mandates of participating institutions
    5. to ensure equity in sharing of the resources,
2. capacity assessment and development to include human, infrastructure, and financial across the sectors, and
3. training of field teams in best practices regarding sample collection and processing at field level [32–34].

Effective coordination of field teams is critical as many of the areas from where the collections are made are remote with no electricity. Due to the rough terrain and impassable roads, especially during the wet season and when the disease transmission is high, a lot of liquid nitrogen may be lost. In view of this, it is important to ensure adequate liquid nitrogen is available to last the period of the field trip. This is critical and assures viability of the parasites from the remote field sites to the laboratories for preparation and permanent storage.

In conclusion, it is possible for institutions to collect, process, store, and maintain biological resources according to their legal mandates in resource-constrained settings. This is only possible through strategic leadership that recognizes the importance of these biological materials to the respective countries and communities from which they are collected. And for organizations requesting for these materials to recognize the efforts and cost of the collection, storage, and maintenance and follow the national and internationally recognized guidelines, rules and regulations regarding the sharing of the same.

Acknowledgements

The Director-General KALRO for financial support and members of staff of the Biotechnology Research Institute-Biochemistry Division and in particular John Ndichu for ensuring regular refilling of stabilate storage dewars with liquid nitrogen.

Author details

Murilla Grace*, Ndung'u Kariuki, Joanna Auma, Purity Gitonga and Thuita John

*Address all correspondence to: gmurilla@yahoo.co.uk

Biotechnology Research Institute, Kenya Agricultural and Livestock Research Organization, Kikuyu, Kenya

References

[1] Murilla, G., et al., Kenya Trypanosomiasis Research Institute cryobank for human and animal trypanosome isolates to support research: opportunities and challenges. PLoS Negl Trop Dis, 2014. 8(5): p. e2747.

[2] Brun, R., et al., The phenomenon of treatment failures in Human African Trypanosomiasis. Trop Med Int Health, 2001. 11: pp. 906–914.

[3] Desquesnes, M. and Dia, M.L., *Trypanosoma vivax*: mechanical transmission of in cattle by one of the most common African tabanids, Atylotusagrestis. Exp Parasitol, 2003. 103(1–2): pp. 35–43.

[4] Hide, G., History of sleeping sickness in East Africa. Amer Soc Microbiol, 1999. 12: pp. 112–125.

[5] Simarro, P., et al., Eliminating human African Trypanosomiasis: where do we stand and what comes next? PLoS Med, 2008. 5(2): p. e55.

[6] WHO, Human African trypanosomiasis (sleeping sickness): epidemiological update. Wkly Epidemiol Rec, 2006. 81: pp. 71–80.

[7] WHO. African trypanosomiasis (sleeping sickness). 2010 [cited 2016 04 July]; Available from: http://www.who.int/trypanosomiasis_african/country/country_situation/en/

[8] Simarro, P., et al., The Atlas of human African trypanosomiasis: a contribution to global mapping of neglected tropical diseases. Int J Health Geogr, 2010. 9: p. 57.

[9] WHO. African trypanosomiasis (sleeping sickness). 2012. http://www.who.int/mediacentre/factsheets/fs259/en/ [cited 04.07.16].

[10] WHO, Control and surveillance of African trypanosomiasis, in WHO technical report series. 1998.

[11] http://www.fao.org/docrep/006/X0413E/X0413E04.htm.

[12] Woo, P., The haematocrit centrifuge technique for the diagnosis of African trypanosomiasis. Acta Tropica, 1970. 27(4): pp. 384–386.

[13] Buscher, P., et al., Improved models of mini anion exchange centrifugation technique (mAECT) and modified single centrifugation (MSC) for sleeping sickness diagnosis and staging. Plos Negl Trop Dis, 2009. 3(11): p. e471.

[14] Herbert, W.J., and W.H.R. Lumsden, A rapid 'matching method' for estimation of host parasitemia. Exp Parasitol, 1976. 40: pp. 427–431.

[15] Cunningham, M.P., W.H. Lumsden, and W.A. Webber, Preservation of viable trypanosomes in lymph tubes at low temperatures. Exp Parasitol, 1963. 14: pp. 280–284.

[16] Dar, F.K., G.S. Lighart, and A.J. Wilson, Cryopreservation of pathogenic African trypanosomes in situ: metacyclics and blood stream forms. J Protozool, 1972. 19(3): pp. 494–497.

[17] Maina, N., et al., Isolation and propagation of *T. b. gambiense* from sleeping sickness patients in S. Sudan. Trans R Soc Trop Med Hyg, 2007. 101: pp. 540–546.

[18] Ndungu K, Gitonga P, Kagira J, Kibugu J, Mulinge M, Munga L, et al. The efficacy of Triladyl 50% in cryopreservation of low concentrations trypanosomes. J Protozoo Resear. 2010;20:27–31.

[19] Baker, J.R., Techniques for the detection of Trypanosomes infections, ed. W.H. Mulligan. 1970, The African Trypanosomiasis. George Allen and Unwin: London. pp. 68–82

[20] Chappuis, F., et al., Options for field diagnosis of human African Trypanosomiasis. Clin Microbiol Rev, 2005. 18: pp. 133–146.

[21] Smith, J.C., et al., Cloning of African trypanosomes in mice immunosuppressed by cyclophosphamide treatment. Amer J Trop Med Hyg, 1982. 31(6): pp. 1098–1102.

[22] Polge, C. and M.A. Soltys, Preservation of trypanosomes in frozen states. Tran R Soc Trop Med Hyg, 1957. 51(6): pp. 519–526.

[23] Otieno, L.H. and N. Darji, Characterization of potentially man infective Trypanosomeabrucei from endemic area of sleeping sickness. Trop Med Parasitol, 1985. 36: pp. 123–124.

[24] Lorne, E.S., ed. Trypanosomiasis a veterinary perspective. 1st ed. 1986, Pergamon Press: England. pp. 235–237

[25] Lumsden WHR. Characterization and nomenclature of trypanosome serodemes and zymodemes: Report of a meeting held in Edinburgh, September 1978. Systematic Parasitology. 1982 4(4):373–6.

[26] Benton, B., Economic impact of onchocerciasis control through the African programme for Onchocerciasis Control. An overview. Ann Trop Med Parasitol, 1998. 92(suppl 1): pp. 33–39.

[27] Hotez, P., et al., The neglected tropical diseases. The ancient afflictions of stigma and poverty and the prospects for their control and elimination. Adv Exp Biol Med, 2006. 582: pp. 22–33.

[28] Ramaiah, K.D., et al., The economic burden of lymphatic filariasis in India. Parasit Today, 2000. 16: p. 251–53.

[29] Molyneux D.H., P.J. Hotez, and A. Fenwick, Rapid-impact Interventions how a policy of integrated control for Africa neglected tropical diseases could benefit the poor. PLos Med, 2005. 2: p. e336.

[30] Kim, A., A. Tandon, and E. Ruiz-Tiben, Cost-benefit analysis of global dracunculiasis eradication campaign, in policy working paper number 1835, T.W. Bank, Editor. 1997: Washington, DC.

[31] http://www.unep.org/newscentre/Default.aspx?DocumentID=26827&ArticleID=35188.

[32] Best Practices for biorepositories; Collection, Storage, Retrieval and Distribution of Biological Materials for Research 2nd ed. International Society for Biological and Environmental Repositories (ISBER). 2008.

[33] NCI Best Practices for Biorepository Resources; Office of Biorepositories and Biospecimen Research; National Cancer Institute, N.I.f.H., Us Department of Health and Human services, 2008.

[34] UNEP, Coordination of Biodiversity-Related Conventions Given a Boost with the Launch of a New Sourcebook. 2015.

Section 2

Cell and Tissue Cryopreservation

Chapter 2

Microscopic Evaluation of Necrotic Cell Death in the Cartilage Destined for Experimental Tracheal Allografts: Lyophilization vs Cryopreservation

Avelina Sotres-Vega, Jaime Villalba-Caloca,
Miguel O. Gaxiola-Gaxiola,
Miriam Mendoza-Santiago,
Juan A. Morales-Tiburcio,
Claudia Hernández-Jiménez and
Francisco Figueroa-Cavero

Additional information is available at the end of the chapter

http://dx.doi.org/10.5772/65014

Abstract

Tracheal replacement remains an important unmet need for patients with extensive lesions. Tracheal allografts treated by cryopreservation and lyophilization have been used as experimental methods for replacing long segments of the trachea. We compare the effect of lyophilization and cryopreservation on the canine tracheal cartilage by microscopic evaluation of necrotic cell death. Canine tracheal segments were rinsed and randomly divided into a control group (G1) and two biopreservation groups: lyophilization (G2) [−70–55°C/10 mmBar] and cryopreservation (G3) [RPMI-1640 + 10%DMSO + 10%SBF, −70°C/−196°C]. After tracheal segments were rehydrated (G2) or thawed (G3), the central ring was obtained from each tracheal segment and processed for histological evaluation with hematoxylin and eosin and for caspase-3 expression by immunohistochemistry. Compared with the control group, chondrocytes without apparent abnormalities, nucleus with karyorrhexis, and caspase-3 expression decreased significantly with the effect of lyophilization and cryopreservation ($p < 0.001$, ANOVA + Tukey, chi-square, Kruskal-Wallis), while a significant decrease in pyknotic nuclei was observed only with the effect of the lyophilization as well as an increase in the nucleus with karyolysis and empty lacunae ($p < 0.001$, ANOVA + Tukey). The mean percentages of normal chondrocytes and empty lacunae were significantly affected by lyophilization compared with cryopreservation ($p < 0.01$, ANOVA + Tukey). Our results strongly suggest that lyophilization has a deleterious impact on the tracheal cartilage.

Keywords: Trachea, Tracheal allograft, Cryopreservation, Lyophilization, Cell death

1. Introduction

End-to-end anastomosis after an extensive tracheal resection is the method of choice for tracheal pathologies such as benign stenosis secondary to post-intubation injury, trauma, congenital, iatrogenic, or neoplasm causes. It is successfully performed in defects affecting up to 50% of the trachea in adults and 33% in infants. More extensive lesions are not amenable to conventional surgical intervention. Tracheal reconstruction after extensive resection is unresolved and remains one of the most important challenges in tracheal surgery [1]. In the search for alternative methods of replacing long tracheal segments, segmental trachea substitution using tracheal segments biopreserved by cryopreservation and lyophilization has been attempted; however, the clinical application of the procedure has been limited due to the fact that contradictory results have been reported [1–3]. A series of interconnected cartilage rings maintain the tubular shape of the trachea, allowing the passage of air. Chondrocytes are the resident cells of the tracheal cartilage. They reside in cavities in the matrix called cartilage lacunae. Chondrocyte connections to each other are crucial for adequate matrix balance and function, determining the tracheal cartilage stiffness of the biopreserved tracheal allograft. The biopreservation of the cartilage must therefore be investigated in basic research models of chondrocyte injuries. Necrosis, or irreversible cell death, is characterized by nuclear swelling, pyknosis, karyorrhexis, karyolysis, and cytoplasmic eosinophilic staining. Dehydration, rehydration, freezing, and thawing result in increased cell death by both necrosis and apoptosis.

Caspase-3 is the primary inducer of cell death by apoptosis. It is implicated in tissue damage due to mechanical ventilation [4], ischemia-reperfusion [5, 6], orotracheal cannulation due to the tidal volume effect [7], and freezing and thawing processes inherent in the cryopreservation of different tissues and cells [1].

The aim of this study was to compare the effect of two preservation methods (lyophilization and cryopreservation) on the canine tracheal cartilage by microscopic evaluation of necrotic cell death.

2. Experimental design

2.1. Material and methods

The protocol was reviewed and approved by the ethics committee of the Instituto Nacional de Enfermedades Respiratorias (INER) ("Ismael Cosio Villegas") and carried out under the Technical Specification for the Care and Use of Laboratory Animals of the Mexican Official

Norm [8] and the Guide for the Care and Use of Laboratory Animals prepared by the US National Institutes of Health [9].

We harvested nine tracheas from dogs weighing 15–30 kg, regardless of sex and age, which were at the end of non-related tracheal research studies. Initial anesthesia was induced by intravenous administration of 0.1 mg/kg xylazine hydrochloride (Bayer, Leverkusen, Germany) and 6 mg/kg propofol (Pisa, Jalisco, Mexico). Anesthetized animals were placed in the supine position, intubated with an endotracheal tube, and placed on mechanical ventilation (Harvard Apparatus and a Vaporizer Isotec 3 Ohmeda). Anesthesia was maintained with 2% isoflurane, FiO_2 100%, tidal volume 15 ml/kg, and respiratory rate 20/min. The neck and the thorax of each animal were shaved and prepared with povidone-iodine solution (EQM, Mexico, DF). A midline cervical to thoracic incision was made. After separating the strap muscles, the trachea was exposed from the cricoid cartilage to the carina and thoroughly dissected. Before harvesting the entire trachea, euthanasia was carried out using an intravenous administration of pentobarbital overdose and 1 mg KCl (Pisa, Jalisco, Mexico). The surrounding tissue was dissected on a cold Mayo table. The trachea was trimmed into seven ring segments and rinsed with a 50% glucose solution (Pisa, Jalisco, Mexico) with 5000 IU heparin/L (Pisa, Jalisco, Mexico) and 0.1 ml/L of antibiotic-antimycotic solution (Sigma, St. Louis, MO, USA). Immediately after washing in glucose solution, the tracheal segments were randomly divided into a control group without preservation (Group 1: nonpreservation tracheal segments) and two biopreservation groups (lyophilization and cryopreservation).

2.2. Lyophilization

Each tracheal segment was mounted on a 13 × 100 mm polypropylene tube (Datalab, Barcelona, Spain), transferred into a Kitasato flask (Pyrex, Corning, USA), and sealed with a polypropylene plug and parafilm (Bemis, Wisconsin, USA). The glass container was placed inside a polystyrene foam box and stored for 24 h at −70°C using a Revco freezer (Thermo Fisher Scientific, Georgia, USA). The Kitasato flask was then introduced into the lyophilizing device (Labconco, Kansas City, MO, USA) at −55°C and 10 mmBar vacuum pressure for 24 h. The tracheal segment was then sealed in airtight double-layered polyethylene bags, sterilized with ethylene oxide (Steri-Vac, Sterilizer/Aerator 5XL, 3 M, USA) at 736 mg ethylene oxide/L air, and stored at room temperature for 30 days. For rehydration, the lyophilized tracheal segment was placed inside a cold glass beaker with saline solution at 4°C for 30 min (Pisa, Jalisco, Mexico). After rehydration, it was removed from the polypropylene tube which was used as a support.

2.3. Cryopreservation

The solutions used for cryopreservation of the tracheal segments were RPMI-1640 media (R8758, Sigma, USA) with 10% dimethyl sulfoxide (D2650, Sigma, USA), 20% fetal bovine serum (16000-044, Gibco, USA), and 0.1 ml antibiotic-antimycotic solution (A5955, Sigma, USA) added per liter of solution. Each tracheal segment was transferred into a cryogenic vial (Nalgene, New York, USA), and the cryopreservation solution was added. Cryogenic vials were placed in a high-density polyethylene vial holder (Thermo Fisher Scientific, Nalgene 5200

Cryo −1°C/min Mr. Frosty Freezing Container, New York, USA) and stored for 24 h at −70°C in a Revco freezer (Thermo Fisher Scientific, Georgia, USA). The cryogenic vials were then stored in liquid nitrogen (–196°C) for 60 days. At the end of cryopreservation, the cryogenic vials were placed in a double boiler at 37°C for 30 min and immediately rinsed for 3 min in saline solution maintained at 4°C (Pisa, Jalisco, Mexico) with three changes of solution.

2.4. Histological and caspase-3 evaluations

Control group (G1): After harvesting, the tracheal segments were rinsed with a 50% glucose solution (Pisa, Jalisco, Mexico) and fixed with 10% buffered formalin for 24 h. Biopreserved groups (G2 and G3): After lyophilization and cryopreservation, tracheal segments were rehydrated (G2) or thawed (G3) and fixed with 10% buffered formalin for 24 h. After tracheal segments were fixed, the central ring was obtained from each tracheal segment, embedded and blocked in paraffin wax (McCormick Scientific, St. Louis, MO, USA), cut into 2 μm and 4 μm thick sections, and mounted on glass slides (Kling-On HIER Slides, Biocare Medical, USA) with a rotary microtome (Reichert, New York, USA). The rings were then processed for histological evaluation with hematoxylin and eosin (Merck, Darmstadt, Germany) and for caspase-3 expression. Histological assessment included the measurement of the percentage (average of four counts) of nucleated chondrocytes, empty lacunae, and cells with pyknosis, karyorrhexis, and karyolysis along the entire tracheal ring (**Figure 1**).

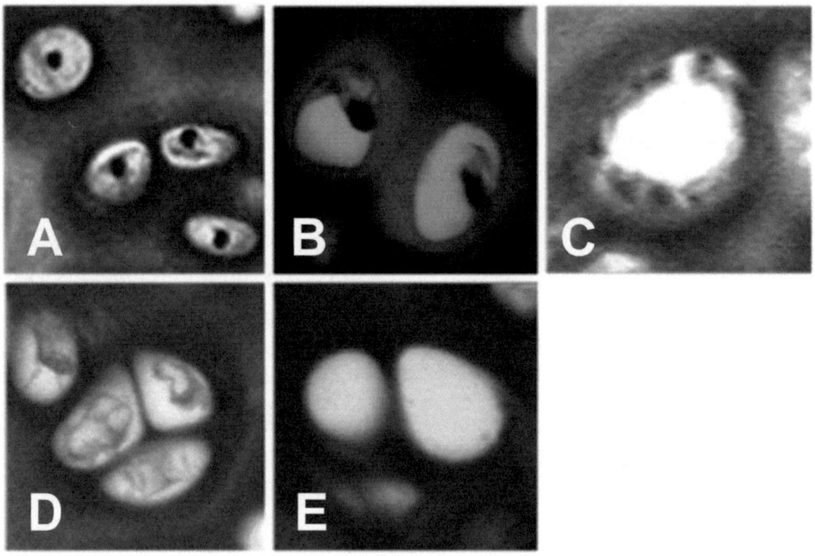

Figure 1. Histological assessment included the measurement of the percentage of nucleated chondrocytes (A), cells with pyknosis (B), karyorrhexis (C), karyolysis (D), and empty lacunae (E) along the entire tracheal ring. Hematoxylin and eosin (40×).

Caspase-3 expression in the cartilage was determined by immunohistochemistry using a polyclonal antibody [Caspase 3 (CPP32) Ab-4, rabbit polyclonal antibody, NeoMarkers, Lab

Vision, RB-1197-P, USA], biotin-avidin-peroxidase system (Vector, California, USA), staining with aminoethylcarbazole (BioGenex, California, USA), and double staining with hematoxylin (Vector, California, USA). Negative and positive controls for caspase-3 expression were counted using an optical microscope (Carl Zeiss, Jena, Germany) and a manual counter (Thermo Fisher Scientific, Waltham, MA, USA). Caspase-3 assessment included measurement of the percentage (average of three counts) of chondrocytes positive for caspase-3 immunoreactivity along the tracheal ring.

Statistical analysis was performed using SPSS 6.1 (Statistical Product and Service Solutions Inc., Chicago, IL, USA). The ANOVA + Tukey test was used to compare. The mean percentages of unaltered chondrocytes, degenerated cells (pyknosis, karyorrhexis, karyolysis), and empty lacunae chi-square and Kruskal-Wallis tests were used to compare caspase-3 expression in the cartilage. A value of $p < 0.05$ was considered statistically significant.

2.5. Results

The mean percentages of unaltered chondrocytes, degenerated cells (pyknosis, karyorrhexis, karyolysis), and empty lacunae obtained over the entire cartilage tracheal ring in all study groups are reported in **Table 1** and illustrated in **Figure 2**.

Nuclei	Normal	Pyknosis	Karyorrhexis	Karyolysis	Empty lacunae
Control	14.73 ± 0.69	26.36 ± 0.88	0.94 ± 0.16	24.02 ± 1.47	33.92 ± 1.34
Post-lyophilization	4.73 ± 0.49	15.78 ± 0.75	0.17 ± 0.05	35.45 ± 2.11	43.86 ± 1.41
Post-cryopreservation	9.15 ± 0.77	24.96 ± 1.41	0.25 ± 0.06	31.29 ± 2.57	34.26 ± 1.70

Table 1. Mean percentage ± standard error of chondrocytes without apparent alterations and degenerated cells.

The lyophilized tracheal segments showed a decreased percentage of unaltered chondrocytes [14.73–4.73 (67.89%)], nuclei with pyknosis [26.36–15.78 (40.14%)], and chondrocytes with karyorrhexis [0.94–0.17 (81.92%)] compared to tracheal segments without lyophilization. The percentage of chondrocytes with karyolysis [24.02–35.45 (47.58%)] and empty lacunae [33.92–43.86 (33.23%)] increased after lyophilization. All changes were statistically significant (control vs post-lyophilization: $p < 0.001$, ANOVA + Tukey).

Cryopreserved tracheal segments showed a decreased percentage of chondrocytes without apparent abnormalities [14.73–9.15 (35.01%)], nuclei with pyknosis [26.36–24.96 (5.32%)], and karyorrhexis [0.94–0.25 (73.41%)], as well as an increase in the percentage of cells with karyolysis [24.02–31.29 (30.26%)] and empty lacunae [33.92–34.26 (1.0%)]. The changes in the percentages of normal chondrocytes and cells with karyorrhexis were significantly different from the control group [control vs post-cryopreservation: chondrocytes unaltered and karyorrhexis ($p < 0.001$), ANOVA + Tukey]. Percentages of nuclei with pyknosis, karyolysis, and empty lacunae were not significantly different when compared to the control group [control vs post-cryopreservation: pyknosis ($p = 0.798$), karyolysis ($p = 0.060$), and empty lacunae ($p = 0.998$, ANOVA + Tukey).

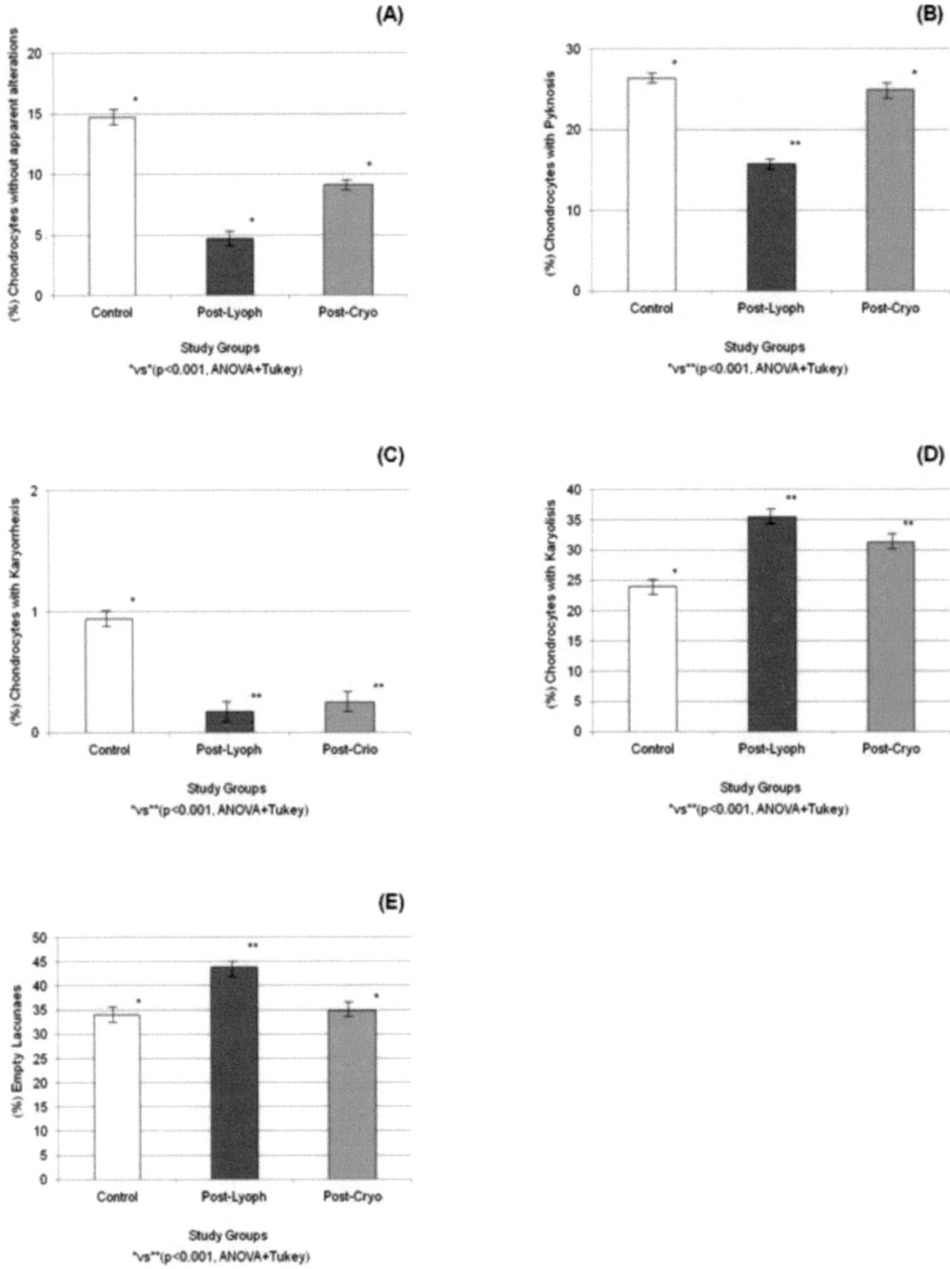

Figure 2. Percentage of chondrocytes without apparent alterations (A), nuclei with pyknosis (B), karyorrhexis (C), and karyolysis (D) before and after biopreservation of the tracheal segments.

The mean percentages of normal chondrocytes, pyknotic nuclei, and empty lacunae were significantly affected by lyophilization compared with cryopreservation [normal chondrocytes, pyknotic cells, and empty lacunae ($p < 0.01$), ANOVA + Tukey].

http://dx.doi.org/10.5772/65014

Only four tracheal segments from the control group were positive for caspase-3 expression in the chondrocytes. This expression was significantly diminished after biopreservation of tracheal segments [control vs post-lyophilization and control vs post-cryopreservation: ($p < 0.01$), chi-square, Kruskal-Wallis] (**Figure 3**).

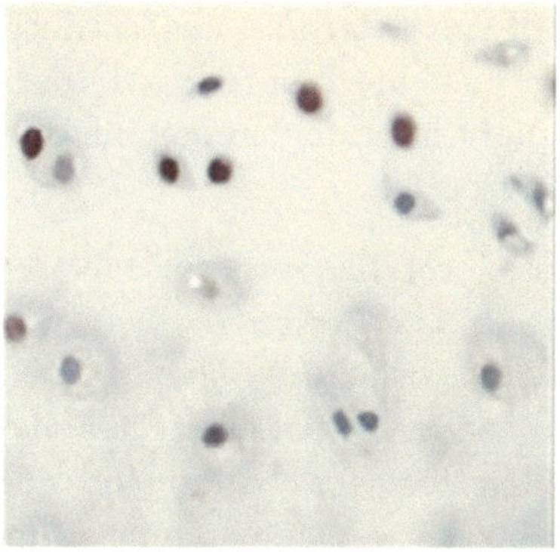

Figure 3. Caspase-3 immunoreactivity (chondrocytes with brown coloration) in one tracheal segment from the control group. Aminoethylcarbazole-hematoxylin staining under ×25 magnification.

3. Discussion

Despite numerous attempts made with synthetic prostheses and tracheal transplants of autologous tissues, none of these alternatives has permitted functional reconstruction due to complications such as devascularization, stenosis, necrosis, dehiscence, infection, immune reactions, and formation of granulation tissue. Reconstruction of long-segment tracheal defects is an important, unresolved clinical problem. Several attempts at replacing the trachea have been made using lyophilized and cryopreserved tracheal segments. We compared the effect of two preservation methods (lyophilization and cryopreservation) on the canine tracheal cartilage by microscopic evaluation of necrotic cell death.

Cell death by means of histological changes and expression of caspase-3 in lyophilized and cryopreserved tracheal cartilage was evaluated. Histology was evaluated by light microscopy. Our results showed that the tracheal cartilage treated with lyophilization or cryopreservation induced significant changes in the chondrocyte integrity. These changes were more severe with lyophilization.

The nucleated chondrocytes decreased significantly with preservation (lyophilization or cryopreservation) compared to the tracheal segments from the control group. However, unaltered chondrocytes were significantly affected by lyophilization compared with cryopreservation. Major cell death could lead to a failure in cartilage matrix turnover because chondrocytes are the only source of matrix component synthesis in the cartilage. If chondrocytes are crucial for adequate matrix balance and function, chondrocyte death could be related to tracheal death.

Chondrocytes reside in cavities in the matrix called cartilage lacunae. In the bone, the standard technique for determining osteonecrosis in clinical pathology remains the identification of empty lacunae [10, 11]. Most osteocyte number studies used the number of lacunae as a reference [11]. Empty osteocyte lacunae rarely occur within the first seven days of bone death, and it may take over 16 weeks for a complete loss of osteocytes from the lacunae [10].

We found that the percentage of empty lacunae increased significantly with lyophilization. This increase was 33% higher than in cryopreserved tracheal segments. This increased percentage of empty lacunae might indicate that lyophilized chondrocytes are prone to die sooner than cryopreserved chondrocytes.

A series of characteristic morphological changes occurs when chondrocytes lose their viability. Necrosis, or irreversible cell death, is characterized by nuclear swelling, pyknosis, karyorrhexis, karyolysis, and cytoplasmic eosinophilic staining [10]. Several studies over the past decade have shown convincing evidence that osteocytes die by apoptosis. During the last stages of programmed cell death, osteocytes break up into apoptotic bodies that may be less than 2–3 μm in diameter. Such bodies are likely to remain undetectable to light microscopy and may well be interpreted as empty lacunae [12]. On the other hand, apoptotic bodies are phagocytosed by neighboring cells or macrophages, thereby preventing the retention of cellular debris in the extracellular space. In the bone, some osteocytes die in situ by apoptosis. This process is brief, and the remnants of apoptotic death are recognizable by conventional light microscopy. Apoptotic bodies can remain as pyknotic nuclei for many months. Eventually, the products of cell death are removed and become undetectable, leaving apparently empty lacunae. Much later, the empty lacunae become filled with mineralized debris and may no longer be visible by light microscopy [13]. Our results indicate that the percentage of chondrocytes with nuclear pyknosis decreased after both lyophilization and cryopreservation. This decrease was statistically significant only in the lyophilized group of tracheal segments, and pyknosis does not necessarily mean necrosis. Coupled with the decrease in the percentage of pyknotic nuclei after preservation, we also found that caspase-3 expression was significantly diminished due to lyophilization or cryopreservation of the tracheal segments. If caspase-3 expression is implicated in tissue damage due to ischemia [5, 6], orotracheal cannulation due to the tidal volume effect [7], and the freezing process inherent in the preservation, it is possible that in lyophilized or cryopreserved tracheal segments, cell death by apoptosis or expression of caspase-3 occurs before biopreservation. The main cell death pathway after rehydrating or thawing tracheal segments would then be necrosis. According to Hedgecock [12], if large numbers of osteocytes undergo apoptosis at the time of tissue harvest, many other osteocytes likely underwent apoptosis previously, becoming shrunken and fragmented and disappearing from their lacunae. Empty lacunae may represent osteocytes that previously died by apoptosis. We also found that lyophilized or cryopreserved tracheal segments showed a significantly decreased percentage of nuclei with karyorrhexis and an increased percentage of nuclei with karyolysis. However, this increase was significant only after lyophilization. Karyolysis is usually preceded by karyorrhexis and occurs primarily as a result of necrosis. In apoptosis, karyorrhexis usually follows after the core is dissolved in apoptotic bodies. The mean percentage of normal or unaltered chondrocytes, pyknotic nuclei, and empty lacunae was significantly affected by lyophilization compared with cryopreservation. Our results therefore strongly suggest that lyophilization has a deleterious impact on the tracheal cartilage and corroborate the findings reported by Lenot et al. [2] and Villalba-Caloca [14].

4. Conclusions

By microscopic evaluation of necrotic cell death, lyophilization has a deleterious impact on the tracheal cartilage.

Author details

Avelina Sotres-Vega[1*], Jaime Villalba-Caloca[1], Miguel O. Gaxiola-Gaxiola[2], Miriam Mendoza-Santiago[1], Juan A. Morales-Tiburcio[1], Claudia Hernández-Jiménez[3] and Francisco Figueroa-Cavero[1]

*Address all correspondence to: avelinasotres@gmail.com

1 Experimental Lung Transplant Unit, National Institute of Respiratory Diseases, "Ismael Cosio Villegas", Mexico City, Mexico

2 Morphology Department, National Institute of Respiratory Diseases, "Ismael Cosio Villegas", Mexico City, Mexico

3 Experimental Surgery Department, National Institute of Respiratory Diseases, "Ismael Cosio Villegas", Mexico City, Mexico

References

[1] Sotres-Vega A, Villalba-Caloca J, Jasso-Victoria R, Olmos-Zíñiga JR, Gaxiola-Gaxiola M, Baltazares-Lipp M, Santibañez-Salgado A. Cryopreserved tracheal grafts: A review of the literature. J Invest Surg 2006;19:125–35.

[2] Lenot B, Macchiarini P, Dulmet E, Weiss M, Dartevelle P. Tracheal allograft replacement. An unsuccessful method. Eur J Cardiothorac Surg 1993;7:648–52.

[3] Haykal S, Salna M, Waddell T, Hofer S. Advances in tracheal reconstruction. Plast Reconstr Sur Glob Open 2014;2:e178; doi: 10.1097/GOX0000000000000097).

[4] Bradley N, Smuder A, Hudson M, Talbert E, Powers S. Cross-talk between the calpain and caspase-3 proteolytic systems in the diaphragm during prolonged mechanical ventilation. Crit Care Med 2012 40: 1857–1863. doi:10.1097/CCM.0b013e318246bb5d.

[5] Petrache I, Fijalkowska I, Medler T, Skirball J, Cruz P, Zhen L, Petrache H, Flotte T, Tuder R. α-1 Antitrypsin inhibits caspase-3 activity, preventing lung endothelial cell apoptosis. Am J Pathol 2006; 169: 1155–1166. doi: 10.2353/ajpath.2006.060058

[6] Hunaid A, Galiñales M. Effect of degree of ischemic injury and reoxygenation time on the myocardial cell death in man: Role of caspases. BMC Physiol 2005;5:1–12.

[7] Hu G, Cai S, Chen Y, Gao F. Effect of different tidal volume ventilation on rat bronchial and alveolar epithelial cell apoptosis. Di Yi Jun Da Xue XueBao 2005;25:508–512.

[8] Official Journal of the Federation. Technical specifications for the production care and use of laboratory animals of the Official Mexican Standard NOM-062-ZOO-1999. Mexico: United States of Mexico (December 6); 1999.

[9] National Institutes of Health U.S.A. Guide for Care and Use of Laboratory Animals. U.S. Eighth Edition. The Guide NRS,http://www.nap.edu/catalog/12910.html; 2011.

[10] Kieser D, Poole T, Jennings M, Parker K, Kieser S, Kieser J, Theis J. The validity of CMFDA and Ethd-1 in the determination of osteocyte viability in the diaphysis of long bones in a sheep model. J Clin Exp Pathol 2014; 4: 201. doi:10.4172/2161-0681.1000201

[11] Parfitt M. Life history of osteocytes: relationship to bone age, bone remodeling, and bone fragility. J Musculoskel Neuron Interact 2002; 2:499–500

[12] Hedgecock N, Hadi T, Chen A, Curtiss S, Martin R, Hazelwood S. Quantitative regional associations between remodeling, modeling, and osteocyte apoptosis and density in rabbit tibial midshafts. Bone 2006;40:627-637, 2007. DOI: http://dx.doi.org/10.1016/j.bone.10.006

[13] Shijing Q, Sudhaker R, Palnitkar S, Parfitt A. Differences in osteocyte and lacunar density between black and white American women. Bone 2006; 38:130–135.

[14] Villalba-Caloca J. Tisular response of lyophilized and cryopreserved tracheal allograft treated with immunosuppresive therapy and topical application of vascular endothelial growth factor. [Doctoral Thesis]. Mexico: National Polytechnic Institute; 2007.

Chapter 3

Isolation and Cryopreservation of Animal Mesenchymal Stromal Cells

Tina Lombardo, Sabrina Renzi, Silvia Dotti,
Stefano Cinotti and Maura Ferrari

Additional information is available at the end of the chapter

http://dx.doi.org/10.5772/64949

Abstract

Scientific progress in cellular and molecular biotechnology has led to the development of advanced therapies, such as gene therapy, cell therapy, and tissue engineering. The application of stem cells as therapeutic agents has been investigated for several years in human medicine and, more recently, the same approach has been considered in the veterinary field as a novel opportunity for the treatment of animal diseases. Mesenchymal stem cell (MSC)-based therapies seem to contribute to the healing process by several mechanisms due to their peculiar biological features. It has been shown that MSCs could effectively differentiate into the required cell type to replace the damaged tissue. Furthermore, due to their autocrine and paracrine secretory activities, these cells are a powerful source of trophic mediators, growth factors, cytokines, and extracellular matrix components. The clinical application of MSCs needs great amounts of cells designed for *in vivo* implantation that can be obtained following their *in vitro* isolation, serial subcultivations, cryopreservation, and thawing. These procedures could determine their feature changes which could interfere with the therapeutic outcome. For these reasons, to preserve MSCs after *in vitro* manipulation for future applications, standardized quality controls and a reliable long-term cryopreservation method are required.

Keywords: mesenchymal stromal cells, cryopreservation, cryoprotectant, regenerative medicine

1. Introduction

The aims of regenerative medicine are to renew cells, regenerate fully functional tissues, and organs or structures that are lost or damaged after disease, injury, or aging [1]. In recent years, the mesenchymal stem cells or multipotent mesenchymal stromal cells (MSCs) have attracted much attention due to their potential use in regenerative medicine and tissue engineering as shown by the main applications described in the literature and the noteworthy progress that has been made toward their better understanding and characterization [2]. Those cells display a significant therapeutic plasticity as reflected by their advantageous characteristics: the ability to enhance tissue renovation, the immunomodulatory, and anti-inflammatory effects [3, 4] and the possibility to be used for both autologous and allogeneic therapies [5]. For these reasons, MSC-based cell therapies have been investigated for several years in human medicine and, more recently, the same approach has been considered in veterinary medicine as a novel potential therapy for animal diseases [6–9]. While most studies using animal models and even small clinical trials have utilized fresh MSC cultured on-site, cryopreservation of MSC is essential to the widespread application of MSC-based therapies. Cryopreservation allows for MSC to be prepared by specialized facilities, in large batches under the application of accepted quality control measures to ensure their safety. Currently, much information concerning the effects of cryopreservation on MSCs is difficult to interpret because MSCs are frequently isolated from different tissue sources and stored for variable periods of time. The capability of MSCs to survive to storage, maintain their phenotype, and differentiate along multiple lineage pathways upon thawing is of paramount importance if they are banked for future therapeutic purposes.

2. Mesenchymal stem cells

MSCs were first described as a specific cell population by Friedenstein's research group in the late 1960s [10]. Previously, stem cell populations were supposed to reside solely in adult tissues with a high turnover rate, such as blood, skin, hair, gastrointestinal epithelium, and bone. Indeed, these cells are present in variable amounts in specific stem cell "niches" (organs), in almost all the body tissues and even if the exact locations of these niches are poorly understood, there is growing evidence suggesting a close relationship with pericytes [11]. Generally, these cells remain in a quiescent state until activated by significant events, such as during tissue repair after injury or following transplantation, to regain tissues' homeostasis [12, 13]. MSCs are undifferentiated, self-renewable, multipotent adult stem cells originated from the mesoderm germ layer during the embryonic development, characterized by the ability to evolve both *in vitro* and *in vivo* along multiple lineage pathways [14]. Furthermore, MSCs have shown evidence of plasticity by trans-differentiating into a broad range of cell types of mesodermal origin (osteocytes, chondrocytes, adipocytes, and myocytes) [15, 16], but also deriving from other germ layers including ectodermal neurons [17] endodermal hepatocytes [18], endothelial cells [11], and cardiomyocytes [19].

3. Properties of MSCs for cellular therapy

Although MSCs first attracted attention due to their ability to differentiate into various cell types, current data suggest that MSCs, as a result of their peculiar biological features, may not only replace damaged tissues, but may be also capable of secreting several bioactive molecules with paracrine and autocrine properties. Such functional secretions of factors are responsible for trophic [20], antiapoptotic, angiogenic, and antiscar effects [21, 22]. MSCs have a further interesting characteristic, related to the capacity to exert immunoregulatory effects on cells of adaptive and innate immunity, such as T and B lymphocytes, dendritic cells, natural killer cells, and monocytes [23]. These immunomodulatory properties that have been extensively demonstrated by several *in vitro* and *in vivo* studies, seem to permit MSC-allogenic transplantation.

4. Sources of MSCs

In veterinary medicine, the first source reported to contain MSCs was the bone marrow (BM) that in the past was also the most widely used [15]. Nevertheless, more recent studies have identified MSCs with similar properties in almost all mammalian tissues such as skeletal muscle [24, 25], tendon [26], skin [27, 28], adipose tissue [29], periosteum [30], synovial membrane [31], dental pulp [32], peripheral blood [33], umbilical cord blood [34], amniotic fluids [35], and cornea [36].

5. Isolation of MSCs from bone marrow

The postnatal bone marrow (BM) has been the most studied tissue as a source of progenitor cells. It contains at least two cell populations: the hematopoietic stem cells (HSCs), located in proximity to the endosteum, and MSCs that surround the trabeculae and blood vessels [37]. HSCs are capable of regenerating the peripheral blood cell lines and the immune system, and the MSCs [38] are capable of giving rise to tissues of each of the three germ layers. Although MSCs derived from BM are easily separated from the nonadherent hematopoietic fraction of cells by culture and adherence to plastic dishes, BM harvest is an invasive and painful surgical procedure that requires the anesthesia and it could be associated with the risk of complications such as hemorrhage, infection, pneumothorax, or pneumopericardium in horses [39, 40]. Moreover, there is only a very low frequency (0.001–0.01%) of MSCs in bone marrow and these numbers decline with the age of the individual.

6. Isolation of MSCs from adipose tissue

The isolation of rat mature adipocytes and adipose tissue progenitor cells was described in literature for the first time by Rodbell [41]. The protocol was based on the fragmentation of adipose tissue into small portions, followed by enzymatic digestion with collagenase type I at 37°C and the subsequent centrifugation to separate the different cell fractions. The obtained supernatant was composed of mature adipocytes and the pellet fraction consisted of the stromal vascular fraction (SVF) components, which comprise a heterogeneous cell population, including circulating blood cells, fibroblasts, pericytes, and endothelial cells, as well as "pre-adipocytes" or adipocyte progenitors. Stem cells and progenitor cells represent about the 3% of all cell populations [42]. Stem cells derived from the adipose tissue (ASCs) represent a purified population of the adherent stem cells present in the adipose tissue, since all other cell types are removed or die with time. Currently, ASC recovery is quick and easy to perform from the subcutaneous adipose tissue, as it could be successfully collected via lipectomy or from the tail base in horses and from the inguinal region [43] or during ovariohisterectomies in dogs and cats. Stem cells derived from the adipose tissue have been increasingly used for cell therapy both in humans and animals [44], either as freshly isolated, SVF cells, or as cultivated ASCs [43]. ASCs proliferate rapidly with a high cellular activity, making them an ideal source to obtain MSCs [45]. The most important advantage of adipose-derived stem cells is their abundance: from 1 g of adipose tissue an average of $0.5–2.0 \times 10^6$ SVF cells can be isolated, which gives 1–10% of stem cell yield [46]; in comparison, MSCs constitute only 0.001–0.01% of BM [15]. When autologous ASCs are used, the adipose tissue is collected 2 or 3 weeks before the treatment and the animal receives the cultivated cells, but long-term cultivation of ASCs before therapeutic use is not recommended, since the cells may lose their progenitor characteristics [47]. The use of allogeneic ASCs has been also performed; since these cells have immunoregulatory properties [48], this approach would allow the use of species-specific allogeneic cryopreserved cells, avoiding the need for collection of tissue from the patient [49].

7. Cryopreservation

Biopreservation has been characterized by a recent rapid growth since advances in cell therapy, stem-cell research, personalized medicine, cell banking, etc. drive the need for optimized storage protocols. Nevertheless, this field still experiences significant issues with the current techniques including suboptimal survival, loss of poststorage cell function, addition of animal components in storage solutions, and activation of cellular stress pathways which can lead to changes in gene expression and protein denaturation [50]. The clinical application of autologous or allogeneic MSCs requires on demand access to a ready off-the-shelf amount of viable therapeutic doses of MSC and therefore necessitates fast availability to cryopreserved MSC stocks. The aim of cryopreservation is to preserve the therapeutic properties of those cells that maintain unaltered the characteristics of the freshly isolated samples, but the freezing and thawing procedures could determine an alteration of the cellular osmosis which can cause cell injury.

7.1. Freezing

The freezing rate is a fundamental factor for all biological systems in the determination of viability following cryopreserved storage. Several studies have shown that successful cryopreservation of cells in suspension needs sufficiently high cooling rates to reach quickly low

http://dx.doi.org/10.5772/64949

temperatures and avoid slow-cooling injury, but low enough cooling rates to decrease the formation of intracellular ice and avoid rapid-cooling injury [51]. The responses to cooling rates are cell-type specific, as distinct cell types have different membrane permeability parameters. The intracellular dynamics during freezing or thawing could be influenced by many factors which influence cell viability after both of these procedures, affecting the therapeutic outcomes. Among these factors, the subtract desegregation stress before cryopreservation of the cells attached to the plastic, the intracellular ice formation during freezing which can compromise the integrity of the cell membranes and, after thawing, the risk of impairing the membrane and altering other cellular functional characteristics can be listed [52]. Currently, there are two procedures to achieve the efficient cryopreservation of MSCs: conventional slow-freezing and vitrification (rapid cooling). Both of these methods may lead to cell damage during loading/unloading of the cryoprotectant agents (CPAs), freezing, and thawing steps. The slow-freezing procedure is the most commonly used cryopreservation technique in clinics and research laboratories today, because it allows the preparation of large amounts of vials at one time. Cryopreservation by vitrification has shown higher cell survival and it has been recognized as a promising strategy for long-term cell banking. Nevertheless, the difficulty to generate a fast enough heating rate to minimize devitrification and recrystallization-induced intracellular ice formation during rewarming is one of the major problems to be overcome. However, the high CPA concentration that is required to achieve vitrification results in osmotic dehydration to cells. For these reasons, new vitrification methods have emerged as alternative techniques, which have shown the ability to significantly reduce cryoinjury. This approach has been improved for the cryopreservation of organized tissues where even extracellular freezing causes several damages. In fact, in a recent study reported by Wang et al. [53], magnetic induction heating of superparamagnetic nanoparticles was successfully applied to enhance rewarming, with promising results of the vitrified human umbilical cord matrix MSC survival.

7.2. Cryoprotectants

In addition to controlling the cooling rates, one of the major challenges to obtain an effective cryopreservation method is the selection of a suitable CPA, which minimizes the damaging effects of freezing. The most commonly employed CPA for cultured mammalian cells is dimethyl sulfoxide (DMSO) solution, because it is cheap and it has a relatively low cell toxicity. DMSO penetrates cell membrane, reduces intracellular ice formation, and prevents cell damage due to dehydration caused by extracellular ice formed during freezing; on the other hand, it can also decrease the survival rate [54, 55] or induce cell differentiation to neuronal-like cells when added to the cell culture medium [56]. The most common cryopreservation medium to store several types of stem cells has become a solution of 10% (v/v) DMSO and up to 90% (v/v) fetal bovine serum (FBS), despite showing disadvantages. To improve this procedure, MSCs have been cryopreserved using both DMSO and FBS free systems, comprising different polymers either alone or in combination with ethylene glycol, 1,2-propylene glycol, trehalose, sucrose, and/or glucose. In contrast to DMSO that penetrates quickly into the cell, the high molecular weight polymers such as polyvinylpyrrolidone, polyethylene glycol, polyethylene oxide, or polyvinyl alcohol are nonpenetrating and seems to act extracellularly (at 10–40% concentrations), with the increasingly high viscosities at low temperature and avoiding that water molecules form ice crystals [52]. In a study reported by Renzi et al. [57], several cryopreservation solutions for MSCs isolated from equine, ovine, rodent bone marrow, and equine adipose tissue were compared: the best results regarding cell viability were obtained using a solution of fetal bovine serum added with 10% of DMSO. Conversely, in a previous study, Ock and Rho [58] reported that the survival and number of colonies formed by porcine MSCs were significantly decreased following short-term storage (less than a month) into liquid nitrogen (−196°C) and the amount of this decrease was inversely proportional to the DMSO concentration. Those data strongly suggest the use of 5% DMSO instead of conventional 10% DMSO for the cryopreservation of porcine MSCs, for minimizing the CPA toxicity on cells. However, slow freezing with reduced concentration of CPAs has gained much interest in order to decrease the effect of the osmotic shock and chemical toxicity. Nevertheless, the commonly used CPAs are highly toxic at 37°C (body temperature) and could not be applied to patients. For this reason, multistep washing is required to completely remove the highly toxic, cell membrane-permeable cryoprotectants from cryopreserved cells for clinical use, though this procedure is often associated with significant loss of precious cells (~10% during each washing step). Therefore, it is important to achieve cell cryopreservation with nontoxic CPAs. Recently, Rao et al. [59] demonstrate that nanoparticle-mediated delivery of trehalose into mammalian cells has great potential for cryopreserving the human primary adipose derived stem cells (hADSCs) and possibly other types of stem cells to facilitate their ready availability for clinical use. In fact, successful results on cryopreservation of hADSCs using only trehalose as cryoprotectant has been achieved with high survival and undamaged function post cryopreservation.

7.3. Thawing and viability assessment

As well as cooling, optimizing the thawing method of frozen MSCs is also important. Furthermore, in clinical transplantation applications the post-thaw viability assessment has shown to be of paramount importance. Several techniques have already been suggested for thawing frozen sample. A procedure of thaw and wash allows to remove DMSO and cell fragments, but may cause cell loss or cellular aggregation during centrifugation. Thaw, dilution, and wash procedure avoids the problem due to the centrifugation, allowing an osmolar equilibration, but the untoward effects of DMSO and cell debris infusion are not prevented. Currently, the standard method for thaw frozen MSCs, either from slow freezing or vitrification, is to warm them rapidly (>100°C/min) in a water bath at 37°C, until all ice crystals disappear. This method generally results in high post-thaw recovery of viable cells without using high-cost equipment, but it is safer to thaw cells using a dry warming procedure, due to the potential microbiological contaminations of the water bath [60]. Literature suggests that rapid thawing rates (>100°C/min) that can prevent damaging ice crystals during recrystallization are optimal choice and generally results in the best post-thaw recovery and viability of cells [61]. High post-thaw viability of MSCs, comparable to those thawed with the standard method, were obtained by Thirumala et al. [62] with a thawing procedure in a controlled-rate freezing/thawing chamber at 10°C/min. For evaluating the cryopreservation outcomes in terms of post-thaw cell quality and quantity, the selection of the correct viability measurement is essential. The most commonly utilized test, owing to its easiness and quickness, is the Trypan blue dye exclusion assay; however, this method has the disadvantage that it generally overestimates the viable population. Several reports suggested that fluorescence dyes are more accurate and reliable indicators of cell viability [63].

8. Microbiological controls

Biosafety assessment of cryopreserved MSCs is necessary to ensure the safe use of the cells prior to clinical applications. Specific tests for the detection of bacteria, yeast, fungi, mycoplasmas, and viruses should be used as a part of routine and regular quality control screening procedures. To detect low levels of contamination, samples from the cell cultures and their products may be inoculated in either liquid tryptic soy broth (TSB) for the detection of aerobes, facultative anaerobes, and fungi, fluid thioglycollate medium (FTM) for the detection of aerobic and anaerobic bacteria, or onto solid (trypticase soy agar, blood agar, Sabouraud's dextrose agar, and malt extract agar) growth media. These inoculated media may be incubated at different temperatures, reflecting conditions for pathogen culture (37°C) and environmental organisms with lower growth temperature optimal (25°C) in microbiological culture incubators, depending on the specific testing standards used. Mycoplasmas competes with the cells for the nutrients in the culture medium, typical signs of contamination consist in a reduction of the rate of cell proliferation, and changes in cellular physiology including gene expression, metabolism, and phenotype. Among the wide variety of techniques that have been developed to detect mycoplasma contamination of cell cultures, Uphoff and Drexler [64] recommended the PCR analysis for the screening, as it considered the most reliable and useful detection method. The presence of viral agents could be evaluated by a panel of tests to detect pathogens and adventitious viruses. Usually, this panel of tests includes: electronic microscopy, reverse transcriptase detection (as a general test for retroviruses), and other tests to find specific agents, depending on the animal species of the sample.

9. Storage of MSCs

MSCs should be preserved without direct exposure to liquid nitrogen, to reduce the risk of pathogenic cross-contamination. This issue enforces the stem cells banks to store materials at vapor phase of liquid nitrogen. However, recent evidence suggests that storage in vapor phase above liquid nitrogen still carries the risk of cross-contamination [65]. Potentially, infective agents may also enter storage directly from the facility atmosphere, contaminated surfaces, or leaking samples, and they can be accumulated in viable condition. Stem cell banks should also maintain secure liquid nitrogen storage equipment in cryogenic tanks monitored by a specific control and alarm system (−196°C), in order to avoid catastrophic loss of cryopreserved samples. Furthermore, proper storage requires the use of cryovials and labeling systems that will withstand the intended storage conditions: labels and bar codes or other printing systems are chosen for extended storage periods.

10. Future perspectives

MSC-based therapy is a promising treatment in repair and regeneration of injured and pathological tissues. Nowadays, even if this innovative therapy in veterinary medicine is still limited, stem cell technology has attracted attention and is a quickly evolving field, among either competitive horses or companion animals, due to the limitations of pharmacological and other current therapeutic strategies. The clinical application of autologous or allogeneic MSCs requires a ready off-the-shelf amount of viable cells that maintain unaltered the characteristics of the freshly isolated samples. Although the long experience of cells' processing facilities, consensus is lacking on a universally accepted method for the effective cryopreservation protocol of MSCs and on the maximum time of cryopreserved storage. For these reasons, even if several successful clinical results have been reported by several groups, the methods of stem cells administration need to be improved and the protocols standardized, before a broad spectrum of clinical applications can be successfully achieved. Currently, the Italian Ministry of Health funded a research project to evaluate the safety and efficacy of animal cryopreserved MSCs for allogenic use. These cells are stored and available at the Italian Biobank of Veterinary Resources of IZSLER (http://www.ibvr.org) and the activity is in progress (data not published).

Author details

Tina Lombardo*, Sabrina Renzi, Silvia Dotti, Stefano Cinotti and Maura Ferrari

*Address all correspondence to: tina.lombardo@izsler.it

Cell Culture Centre, Experimental Zooprophylactic Institute of Lombardia and Emilia Romagna (IZSLER), Brescia, Italy

References

[1] Brockes JP, Kumar A. Appendage regeneration in adult vertebrates and implications for regenerative medicine. Science. 2005;310(5756):1919–1923. DOI: 10.1126/science.1115200.

[2] Perez-Merino EM, Uson-Casaus JM, Zaragoza-Bayle C, Duque-Carrasco J, Marinas-Pardo L, Hermida-Prieto M, Barrera-Chacon R, Gualtieri M. Safety and efficacy of allogeneic adipose tissue-derived mesenchymal stem cells for treatment of dogs with

inflammatory bowel disease: clinical and laboratory outcomes. The Veterinary Journal. 2015;206:385–390. DOI: 10.1016/j.tvjl.2015.07.023.

[3] Jorgensen C, Noel D. Mesenchymal stem cells in osteoarticular diseases. Regenerative Medicine. 2011;6:44–51. DOI: 10.2217/rme.11.80.

[4] Gimble JM, Bunnell BA, Frazier T, Rowan B, Shah F, Thomas-Porch C, Wu X. Adipose-derived stromal/stem cells: a primer. Organogenesis. 2013;9:3–10. DOI: 10.4161/ org. 24279.

[5] Pavyde E, Maciulaitis R, Mauricas M, Sudzius G, Didziokiene EI, Laurinavicius A, Sutkeviciene N, Stankevicius E, Maciulaitis J, Usas A. Skeletal muscle-derived stem/progenitor cells: a potential strategy for the treatment of acute kidney injury. Stem Cells International. 2016, Article ID 9618480, 13 pp., http://dx.doi.org/ 10.1155/2016/9618480.

[6] Lopez MJ, Jarazo J. State of the art: stem cells in equine regenerative medicine. Equine Veterinary Journal. 2015;47:145–154. DOI: 10.1111/evj.12311.

[7] Kim Y, Lee SH, Kim WH, Kweon Oh-K. Transplantation of adipose derived mesenchymal stem cells for acute thoracolumbar disc disease with no deep pain perception in dogs. Journal of Veterinary Science. 2016;17(1):123–126. DOI: 10.4142/jvs. 2016.17.1.123.

[8] Webb TL, Webb CB. Stem cell therapy in cats with chronic enteropathy: a proof-of-concept study. Journal of Feline Medicine and Surgery. 2015;17:901–908. DOI: 10.1177/1098612X14561105.

[9] Trzil JE, Masseau I, Webb TL, Chang CH, Dodam JR, Liu H, Quimby JM, Dow SW, Reinero CR. Intravenous adipose-derived mesenchymal stem cell therapy for the treatment of feline asthma: a pilot study. Journal of Feline Medicine and Surgery. Sep. 17, 2015:1–10 DOI: 10.1177/1098612X15604351.

[10] Friedenstein AJ, Petrakova KV, Kurolesova AI, Frolova GP. Heterotopic of bone marrow. Analysis of precursor cells for osteogenic and hematopoietic tissues. Transplantation. 1968;6(2):230–247.

[11] Farrington-Rock C, Crofts NJ, Doherty MJ, Ashton BA, Griffin-Jones C, Canfield AE. Chondrogenic and adipogenic potential of microvascular pericytes. Circulation. 2004;110:2226–2232. DOI: 10.1161/01.CIR.0000144457.55518.E5.

[12] da Silva Meirelles L, Chagastelles PC, Nardi NB. Mesenchymal stem cells reside in virtually all post-natal organs and tissues. Journal of Cell Science. 2006;119(11):2204–2213. DOI: 10.1242/jcs.02932.

[13] Woo D-H, Hwang HS, Shim JH. Comparison of adult stem cells derived from multiple stem cell niches. Biotechnology Letters. 2016;38:751–759. DOI: 10.1007/ s10529-016-2050-2.

[14] Salem HK, Thiemermann C. Mesenchymal stromal cells: current understanding and clinical status. Stem Cells. 2010;28(3):585–596. DOI: 10.1002/stem.269.

[15] Pittenger MF, Mackay AM, Beck SC, Jaiswal RK, Douglas R, Mosca JD, Moorman MA, Simonetti DW, Craig S, Marshak DR. Multilineage potential of adult human mesenchymal stem cells. Science. 1999;284:143–147. DOI: 10.1126/science.284.5411.143.

[16] Larsen S, Lewis ID. Potential therapeutic applications of mesenchymal stromal cells. Pathology. 2011;43:592–604. DOI: 10.1097/PAT.0b013e32834ab72d.

[17] Taran R, Mamidi MK, Singh G, Dutta S, Parhar IS, John JP, Bhonde R, Pal R, Das AK. In vitro and in vivo neurogenic potential of mesenchymal stem cells isolated from different sources. Journal of Bioscience. 2014;39:157–169. DOI: 10.1007/s12038-013-9409-5.

[18] Najimi M, Khuu DN, Lysy PA, Jazouli N, Abarca J, Sempoux C, Sokal EM. Adult derived human liver mesenchymal-like cells as a potential progenitor reservoir of hepatocytes? Cell Transplantation. 2007;16:717–728. DOI: http://dx.doi.org/10.3727/000000007783465154.

[19] Carvalho PH, Daibert AP, Monteiro BS, Okano BS, Carvalho JL, Cunha DN, Favarato LS, Pereira VG, Augusto LE, Del Carlo RJ. Differentiation of adipose tissue-derived mesenchymal stem cells into cardiomyocytes. Arquivos Brasileiros de Cardiologia. 2013;100(1):82–89.

[20] Figueroa FE, Carrión F, Villanueva S, Khoury M. Mesenchymal stem cell treatment for autoimmune diseases: a critical review. Biological Research. 2012;45:269–277. DOI: 10.4067/S0716-97602012000300008.

[21] Caplan AI, Dennis JE. Mesenchymal stem cells as trophic mediators. Journal of Cellular Biochemistry. 2006;98:1076–1084. DOI: 10.1002/jcb.20886.

[22] Caplan AI, Correa D. The MSC: an injury drugstore. Cell Stem Cell. 2011;9:11–15. DOI: 10.1016/j.stem.2011.06.008.

[23] Menard C, Tarte K. Immunoregulatory properties of clinical grade mesenchymal stromal cells: evidence, uncertainties, and clinical application. Stem Cell Research & Therapy. 2013;4(3):64. DOI: 10.1186/scrt214.

[24] Dodson MV, Hausman GJ, Guan L, Du M, Rasmussen TP, Poulos SP, Mir P, Bergen WG, Fernyhough ME, McFarland DC, Rhoads RP, Soret B, Reecy JM, Velleman SG, Jiang Z. Skeletal muscle stem cells from animals. I. Basic cell biology. The International Journal of Biological Sciences. 2010;6:465–474. DOI: 10.7150/ijbs.6.465.

[25] Meligy FY, Shigemura K, Behnsawy HM, Fujisawa M, Kawabata M, Shirakawa T. The efficiency of in vitro isolation and myogenic differentiation of MSCs derived from adipose connective tissue, bone marrow, and skeletal muscle tissue. In Vitro Cellular and Developmental Biology-Animal. 2012;48(4):203–215. DOI: 10.1007/s11626-012-9488-x.

[26] Radtke CL, Nino-Fong R, Esparza Gonzalez BP, Stryhn H, McDuffee LA. Characterization and osteogenic potential of equine muscle tissue- and periosteal tissue-derived mesenchymal stem cells in comparison with bone marrow- and adipose tissue-derived mesenchymal stem cells. American Journal of Veterinary Research. 2013;74:790–800. DOI: 10.2460/ajvr.74.5.790.

[27] Shim JH, Kang HH, Lee TR, Shin DW. Enrichment and characterization of human dermal stem/progenitor cells using collagen type IV. Journal of Dermatological Science. 2012;67:202–205. DOI: 10.1016/j.jdermsci.2012.06.007.

[28] Toma JG, Akhavan M, Fernandes KJ, Barnabe-Heider F, Sadikot A, Kaplan DR, Miller FD. Isolation of multipotent adult stem cells from the dermis of mammalian skin. Nature Cell Biology. 2001;3:778–784. DOI: 10.1038/ncb0901-778.

[29] Gimble JM, Katz AJ, Bunnell BA. Adipose-derived stem cells for regenerative medicine. Circulation Research. 2007;100:1249–1260. DOI: 10.1161/01.RES.0000265074.83288.09.

[30] Stewart AA, Barrett JG, Byron CR, Yates AC, Durgam SS, Evans RB, Stewart MC. Comparison of equine tendon-, muscle-, and bone marrow-derived cells cultured on tendon matrix. American Journal of Veterinary Research. 2009;70:750–757. DOI: 10.2460/ajvr.70.6.750.

[31] De Bari C, De Accio FD, Tylzanowski P, Luyten FP. Multipotent mesenchymal stem cells from adult human synovial membrane. Arthritis & Rheumatism. 2001;44:1928–1942. DOI: 10.1002/1529-0131(200108)44:8<1928::AID-ART331>3.0.CO;2-P.

[32] Mead B, Logan A, Berry M, Leadbeater W, Scheven BA. Concise review: dental pulp stem cells: a novel cell therapy for retinal and central nervous system repair. Stem Cells. Jun. 7, 2016:1–7. DOI: 10.1002/stem.2398.

[33] Villaron EM, Almeida J, Lopez-Holgado N, Alcoceba M, Sanchez-Abarca LI, Sanchez-Guijo FM, Alberca M, Perez-Simon JA, San Miguel JF, Del Canizo MC. Mesenchymal stem cells are present in peripheral blood and can engraft after allogeneic hematopoietic stem cell transplantation. Haematologica. 2004;89:1421–1427.

[34] Kern S, Eichler H, Stoeve J, Klüter H, Bieback K. Comparative analysis of mesenchymal stem cells from bone marrow, umbilical cord blood, or adipose tissue. Stem Cells. 2006;24(5):1294–1301. DOI: 10.1634/stemcells.2005-0342.

[35] Iacono E, Cunto M, Zambelli D, Ricci F, Tazzari PL, Merlo B. Could fetal fluid and membranes be an alternative source for mesenchymal stem cells (MSCs) in the feline species? A preliminary study. Veterinary Research Communications. 2012;36(2):107–118. DOI: 10.1007/s11259-012-9520-3.

[36] Moriyama H, Kasashima Y, KuwanoA, Wada S. Anatomical location and culture of equine corneal epithelial stem cells. Veterinary Ophthalmology. 2014;17:106–112. DOI: 10.1111/vop.12050.

[37] Bianco P, Riminucci M, Gronthos S, Robey PG. Bone marrow stromal stem cells: nature, biology, and potential applications. Stem Cells. 2001;19:180–192. DOI: 10.1634/stemcells.19-3-180.

[38] Deans RJ, Moseley AB. Mesenchymal stem cells: biology and potential clinical uses. Experimental Hematology. 2000;28:875–884. DOI: 10.1016/S0301-472X(00)00482-3.

[39] Vidal MA, Kilroy GE, Lopez MJ, Johnson JR, Moore RM, Gimble JM. Characterization of equine adipose tissue-derived stromal cells: adipogenic and osteogenic capacity and comparison with bone marrow-derived mesenchymal stromal cells. Veterinary Surgery. 2007;36:613–622. DOI: 10.1111/j.1532-950X.2007.00313.x.

[40] Peters AE, Watts AE. Biopsy needle advancement during bone marrow aspiration increases mesenchymal stem cell concentration. Frontiers in Veterinary Science. 2016 Mar 14;3:23. DOI: 10.3389/fvets.2016.00023. eCollection 2016.

[41] Rodbell M. Metabolism of isolated fat cells. I. Effects of hormones on glucose metabolism and lipolysis. Journal of Biological Chemistry. 1964;239:375–380.

[42] Tsuji W, Rubin JP, Marra KG. Adipose-derived stem cells: implications in tissue regeneration. World Journal of Stem Cells. 2014;6:312–321. DOI: 10.4252/wjsc.v6.i3.312.

[43] Marx C, Silveira MD, Nardi NB. Adipose-derived stem cells in veterinary medicine: characterization and therapeutic applications. Stem Cells and Development. 2015;24(7): 803–813. DOI: 10.1089/scd.2014.0407.

[44] Del Bue M, Riccò S, Ramoni R, Conti V, Gnudi G, Grolli S. Equine adipose-tissue derived mesenchymal stem cells and platelet concentrates: their association in vitro and in vivo. Veterinary Research Communication. 2008;32(1):S51–S55. DOI: 10.1007/s11259-008-9093-3.

[45] Minonzio G, Corazza M, Mariotta L, Gola M, Zanzi M, Gandolfi E, De Fazio D, Soldati G. Frozen adipose derived mesenchymal stem cells maintain high capability to grow and differentiate. Cryobiology. 2014;69(2):211–216. DOI: 10.1016/j.cryobiol.2014.07.005.

[46] Bear PC, Kuci S, Krause M, Kuçi Z, Zielen S, Geiger H, Bader P, Schubert R. Comprehensive phenotype characterization of human adipose-derived stromal/stem cells and their subsets by a high throughput technology. Stem Cells and Development. 2013;22(2):330–339. DOI: 10.1089/scd.2012.0346.

[47] Geissler S, Textor M, Kühnisch J, Könnig D, Klein O, Ode A, Pfitzner T, Adjaye J, Kasper G, Duda GN. Functional comparison of chronological and in vitro aging: differential role of the cytoskeleton and mitochondria in mesenchymal stromal cells. PLoS One. 2012;7(12):e52700. DOI: 10.1371/journal.pone.0052700.

[48] Carrade DD, Borjesson DL. Immunomodulation by mesenchymal stem cells in veterinary species. Comparative Medicine. 2013;63:207–217.

[49] Park SA, CM Reilly, JA Wood, DJ Chung, DD Carrade, SL Deremer, RL Seraphin, KC Clark, AL Zwingenberger, Borjesson DL, Hayashi K, Russell P, Murphy CJ. Safety and

immunomodulatory effects of allogeneic canine adipose-derived mesenchymal stromal cells transplanted into the region of the lacrimal gland, the gland of the third eyelid and the knee joint. Cytotherapy. 2013;15(12):1498–1510. DOI: 10.1016/j.jcyt.2013.06.009.

[50] Tsonev LI, Hirsh AG. Fluorescence ratio intrinsic basis states analysis: a novel approach to monitor and analyze protein unfolding by fluorescence. Journal of Biochemical and Biophysical Methods. 2000;45:1–21. DOI:10.1016/S0165-022X(00)00070-1.

[51] Marquez-Curtis LA, Janowska-Wieczorek A, McGann LE, Elliott JAW. Mesenchymal stromal cells derived from various tissues: biological, clinical and cryopreservation aspects. Cryobiology. 2015;71(2):181–197. DOI: 10.1016/j.cryobiol.2015.07.003.

[52] Goh BC, Thirumala S, Kilroy G, Devireddy RV, Gimble JM. Cryopreservation characteristics of adipose-derived stem cells: maintenance of differentiation potential and viability. Journal of Tissue Engineering and Regenerative Medicine. 2007;1(4):322–324. DOI: 10.1002/term.35.

[53] Wang J, Zhao G, Zhang Z, Xu X, He X. Magnetic induction heating of superparamagnetic nanoparticles during rewarming augments the recovery of hUCM-MSCs cryopreserved by vitrification. Acta Biomaterialia. 2016;33:264–274. DOI: 10.1016/j.actbio.2016.01.026.

[54] Fuller BJ. Cryoprotectants: the essential antifreezes to protect life in the frozen state. Cryo Letters. 2004;25(6):375–388.

[55] Meryman HT. Cryopreservation of living cells: principles and practice. Transfusion. 2007;47(5):935–945. DOI: 10.1111/j.1537-2995.2007.01212.x.

[56] Woodbury D, Schwarz EJ, Prockop DJ, Black IB. Adult rat and human bone marrow stromal cells differentiate into neurons. Journal of Neuroscience Research. 2000;61:364–370. DOI: 10.1002/1097-4547(20000815)61:4<364::AID-JNR2>3.0.CO;2-C.

[57] Renzi S, Lombardo T, Dotti S, Dessì SS, De Blasio P, Ferrari M. Mesenchymal stromal cell cryopreservation. Biopreservation and Biobanking. 2012;10(3):276–281. DOI: 10.1089/bio.2012.0005.

[58] Ock S-A, Rho G-J. Effect of dimethyl sulfoxide (DMSO) on cryopreservation of porcine mesenchymal stem cells (pMSCs). Cell Transplantation. 2011;20:1231–1239. DOI: http://dx.doi.org/10.3727/096368910X552835.

[59] Rao W, Huang H, Wang H, Zhao S, Dumbleton J, Zhao G, He X. Nanoparticle-mediated intracellular delivery enables cryopreservation of human adipose-derived stem cells using trehalose as the sole cryoprotectant. ACS Applied Materials & Interfaces. 2015;7(8):5017–5028. DOI: 10.1021/acsami.5b00655.

[60] Thirumala S, Goebel WS, Woods EJ. Clinical grade adult stem cell banking. Organogenesis. 2009;5(3):143–154.

[61] Röllig C, Babatz J, Wagner I, Maiwald A, Schwarze V, Ehninger G, Bornhäuser M. Thawing of cryopreserved mobilized peripheral blood-comparison between waterbath and dry warming device. Cytotherapy. 2002;4(6):551–555.

[62] Thirumala S, Zvonic S, Floyd E, Gimble JM, Devireddy RV. Effect of various freezing parameters on the immediate post-thaw membrane integrity of adipose tissue derived adult stem cells. Biotechnology Progress. 2005;21(5):1511–1524. DOI: 10.1021/bp050007q.

[63] Altman SA, Randers L, Rao G. Comparison of Trypan blue dye exclusion and fluorometric assays for mammalian cell viability determinations. Biotechnology Progress. 1993;9(6):671–672. DOI: 10.1021/bp00024a017.

[64] Uphoff CC, Drexler HG. Detection of mycoplasma contaminations. Methods in Molecular Biology. 2013;946:1–13. DOI: 10.1007/978-1-62703-128-8_1.

[65] Grout BW, Morris GJ. Contaminated liquid nitrogen vapor as a risk factor in pathogen transfer. Theriogenology. 2009;71(7):1079–1082. DOI: 10.1016/j.theriogenology.2008.12.011.

Section 3

Sperm Cryopreservation

Chapter 4

Semen Cryopreservation in Brazilian Freshwater Fish: Advances and Main Challenges

Raycon R.F. Garcia, Danilo P. Streit Jr,
Elsa Cabrita and Leandro C. Godoy

Additional information is available at the end of the chapter

http://dx.doi.org/10.5772/65037

Abstract

Studies on semen cryopreservation in Brazilian freshwater fish have been growing in number of publications and investigated species. Despite this apparent increase in research, standardization of cryoprotocols is still missing, making it clear that the grounds on the quality of cryopreserved semen has not yet reached a level that guarantee satisfactory results for its replication. This chapter aims to make a critical and reflective analysis on the ways cryopreservation of freshwater fish semen has been conducted in Brazil. The difficulties in standardizing protocols, broodstock, and selection of genetically superior animals; the barriers in transferring technology from laboratory benches to the field and make feasible the use of cryopreserved semen on a commercial scale; the formation of germplasm banks and the responsible use of cryopreserved material are also discussed. We have no intention to point out the successes and mistakes that may have been committed in pursuing development of cryopreservation protocols, but a reflection on the future directions considering what should be pondered on this subject with objectivity and scientific consolidation.

Keywords: Brazilian freshwater species, sperm cryopreservation, postthaw quality evaluation, cryobanking, aquaculture, fish conservation

1. Introduction

The studies on semen cryopreservation from Brazilian freshwater fish had its beginning in the 1980s with *Prochilodus scrofa* = *P. lineatus* e *Salminus maxillosus* = *S. brasiliensis* species [1]. Basically, the freezing solution used was composed by dimethylsulfoxide (DMSO) as permeating and glucose as nonpermeating cryoprotectant. Other fish species have been studied at that period, such as *Rhamdia hilarii* = *R. quelen* [2] and *Leporinus silvestrii* [3].

The 1990s began with the studies focused on the species target as potential for production in the fish farming scenario at that time, the *Piaractus mesopotamicus* [4, 5]. However, during this decade, there were virtually no further researches in the area and only in 1999 a study with *Colossoma macropomum* [6] was published. However, the study considered the major milestone in fish semen cryopreservation in Brazil was carried out by Carolsfeld et al. [7], who published a series of compiled results describing the freezing protocols for native species, most of them still being used nowadays. Currently, around 20 Brazilian freshwater species have their semen cryopreserved, some even have different protocols [8], or small variations from the protocols described by Carolsfeld et al. [7].

The studies that guided the composition of a future protocol for fish semen freezing produced a lack of originality, resulting in little progress to date. Some factors have contributed to this scenario, such as: the small number of researchers working in the area; restriction of modern equipment; reduced scientific exchange with international groups and even the lack of criteria for distribution of public resources that result in numerous similar publications without significant progress.

This chapter is not intended to point out the successes and mistakes that may have been committed in the pursuit of development of cryopreservation protocols for fish semen in Brazil, but a reflection on the future directions considering what should be pondered on this subject with objectivity and scientific consolidation.

2. Extender solutions

The extenders are used to meet several requirements necessary for sperm survival during cryostorage. Its function is to provide a favorable microenvironment to maintain the viability of sperm cells, allowing the addition of energy sources (lipids, carbohydrates, and metabolites) to support the cellular metabolism, control of pH and osmolality, and prevention of bacterial growth [9].

Since many factors can influence on semen quality parameters, the extender solution must be carefully formulated [10]. The first extender mediums used for South American fish semen were composed only by 0.8% NaCl [1]. Thereafter, an improvement in the results was noticed when glucose was added to the mediums, jointly with egg yolk or powder milk [7].

Egg yolk is still frequently used as a component of extender solutions. According to Watson [11], due to the low-density lipoprotein (LDL) fraction in its composition, the egg yolk may provide a better stabilization of the sperm membrane by passing the cryopreservation stress, reducing injuries, and the thermal shock. The stabilizing mechanism of the sperm membrane by LDL possibly occurs because these compounds attract the molecules of cholesterol present in seminal plasma, thus preventing cholesterol binds to the phospholipids of the sperm membrane, avoiding their destabilization [12]. In *C. macropomum*, the addition of egg yolk in the extender improved sperm motility rate [13].

The composition of the extenders used for preservation of semen from South American fish is quite varied and may be based on salts, glucose, or a bit more complex composition, which includes the addition of energy substrates and antibiotics [14, 15].

The extenders used for the South American fish species are still based on solutions formulated for mammal's semen. Currently, the BTS™ (Beltsville Thawing Solution—Minitub), which is formulated for swine semen is the basic solution that has been used in the composition of extenders [16–19]. Similarly, ACP-104™ (ACP Biotechnology—UECE), which was originally formulated for goats semen was also tested in some fish species [15, 17, 20, 21]. However, seminal viability results obtained in the literature are very variable, indicating that there is a need to seek efforts to formulate specific extenders, based on biochemical composition of seminal plasma from each fish species. Knowledge on the biochemical characteristics of the seminal plasma is essential to understand the spermatozoa requirements, assisting in the preparation of appropriate extender solutions for both short and long-term cryostorage.

3. Cryoprotectants

For performing cryopreservation, either by slow freezing or ultrarapid techniques, the use of permeating and nonpermeating cryoprotectant agents (CPAs) is required [22]. However, CPAs can be very toxic depending on the concentration used and cell exposure to them before freezing should be controlled [22].

The nonpermeable CPA most used by Brazilian researchers for semen from scale fish species is dimethylsulfoxide (DMSO) [8], whereas for nonscale fish species methanol is the most commonly used CPA [7, 23]. In the last years other CPAs have also been tested, such as methyl glycol, glycerol, ethylene glycol, and dimethylformamide (DMF) [8]. We understand that testing new CPAs should be encouraged in research. However, it is important to consider some aspects and thus avoid wasting of time and money. DMSO is no doubt the CPA that has shown the best results to date. We must understand very well the chemical and physical behaviors of a given reagent at low temperatures as well as its permeability and toxicity to sperm cells to be worth DMSO's replacement.

Glucose, as a low molecular weight nonpermeable CPA has been widely used in most protocols in Brazil. However, it is still quite common to observe its use for the formulation of cryoprotectant solutions with egg yolk [8]. Among the cryoprotective effects of glucose on sperm cells are dehydration before cooling, leading to less intracellular ice crystal formation, the increase of effective viscosity of the media, and serving as a protector to membrane integrity. Importantly, glucose is a component commonly added to the cryoprotectant solution and its function as an energy supplier to the spermatozoon and/or cryoprotective action will depend on the concentration used.

We consider as a decisive factor for the composition of the cryomedium not only the choice of cryoprotectants, but also their chemical quality. Although it appears to be negligible, this fact is crucial for a proper and successful protocol. There has been a significant improvement in the supply of chemical reagents, but the delivery time and costs remain a limiting in some parts of the country.

The large number of farmed fish species nowadays in Brazil may be one of the causes of using a high range of different CPAs. However, a standardization of cryoprotocols within a given species should be searched.

4. Freezing and thawing rates

Freezing and thawing rates are determining factors in a gamete cryopreservation protocol. A standard set for the protocols adopted for the South American native fish species is something lacking. In the published studies, we can observe different freezing and thawing rates for different species and even for the same species. Surely, this implies difficulty of adopting a technology that could be replicated and used efficiently by a farmer or even a trained technician. The use of a dry shipper has been established as a common method in research with semen cryopreservation in Brazil [8], since is considered a safe and practical method [24]. Moreover, the dry-shipper container used in research produces similar freezing rates and storage temperatures [13, 25, 26]. However, it is in the thawing rates that the largest variations are noticed, since different temperatures and thawing times are used [8]. For example, Velasco-Santamaría et al. [27] tested 0.5, 1.8, 2.5, and 4.0 ml straws for freezing of *Brycon amazonicus* semen and the results indicated that an increased in thawing temperature from 35 to 80°C for 10 or 90 s could influence sperm motility and fertilization rate. However, for *P. lineatus* the thawing temperature may vary from 30 to 60°C for 8 or 16 s, respectively, without major changes in motility rate when using 0.5 ml straws [18]. For *C. macropomum* semen, a thawing temperature of 45°C for 5–8 s [20, 24] when using 0.25 ml straw and for 0.5 ml straw 37°C/30 s or 60°C/8 s was used [13, 21]. However, at larger scale using 1.6 or 4.5 ml straws these rates can be of 60°C for 90 s [25].

Note that the size of the straw can affect the quality of cryopreserved semen [28] as well as the temperature used for thawing, since high temperatures can denature enzymes and proteins of sperm cells [29]. Therefore, it is necessary to search for a balance between thawing rate and especially size of the container, to be able to maximize the use of cryopreserved semen and have a product in quality and quantity [25].

5. Activating solutions

By having a limited supply of ATP accumulated in the mitochondria [30], the motility duration in freshwater fish spermatozoa is greatly reduced. Motility is a variable that significantly alter the sperm's ability to fertilize the oocyte [31]. Therefore, the use of balanced osmotic media with substrates that can provide energy to the sperm is desirable [23, 32, 33]. The activating solution may also assist in the reduction of osmotic shock, which is a causative effect of damage in sperm cells [34, 35].

Examples of the use of activating solutions in Brazilian research can be found for *R. quelen* semen [23] using a solution composed of fructose, and for *Pseudoplatystoma corruscans* semen [7] using 1% $NaCHO_3$ solution. For scale fish, the use of solutions based on NaCl (0.29%) and $NaHCO_3$ (1%) are shown to have better results [8].

It is noticed that the use of activating solutions is still not as common in Brazil and the previous studies are focused on the use of solutions based on salts or sugars. Fundamentally, future Brazilian researches should explore the biochemical composition of seminal plasma and ovarian fluid. This knowledge will be instrumental in building a successful activating solution.

6. Artificial fertilization

Most papers on seminal cryopreservation from South American fish do not have the data of fertilization rates using cryopreserved semen [8]. The final validation of applying this technique necessarily culminates with the results of fertilization and hatching rates. Nevertheless, to reach this goal some steps must comply in order to create a favorable environment allowing the encounter of the gametes to occur in an efficient and safe manner. In Brazil, in general the most basic aspects involving fish gametes manipulation are still neglected on fish farms. There are preeminent need to establish guidelines for best practices in fish handling during gametes collection and all subsequent semen handling until freezing.

Generally, fertilization rates are lower when using cryopreserved semen, which is a function of reduced quality of sperm cells. Regarding the lack of studies assessing the fertilization and hatching rates, we can relate: the difficulty of having oocytes available for carrying out the tests considering the maintenance costs of a broodstock; experiments poorly designed and lack of methodology standardization in the studies. It is noteworthy that even a motile sperm is no guarantee of fertilization. From more elaborated analysis, discussed in the next topic, it is possible to check for sperm membrane or DNA damage and how such damage can affect their fertilizing capacity.

Indeed, for development of a commercial protocol using cryopreserved semen in Brazil, even on a small scale, an evolution in the control of the above-mentioned processes will be required. If the ultimate goal is to get a good hatching rate (live larvae), that is directly related to the quality of biological material to be cryopreserved, asepsis of facilities and equipment and an efficient cryopreservation protocol. We must consider that the quality of oocytes should be excellent, ensuring that this is not a factor that negatively interfere to the success of the procedure. Are there any criteria for assessing the spawning quality of fish species studied in Brazil? The answer is no, there are not! Nowadays, there is no effective technique to evaluate the quality of oocytes in fish farming in a practical and affordable way.

The inseminating dose (spermatozoa/oocyte ratio) is an important feature that starts to be focused in some studies [36–38]. The entire control of the artificial fertilization environment and optimizing the use of inseminating dose should be the goal for the near future.

7. Main methods used to evaluate cryopreserved semen

7.1. Sperm quality evaluation in fish

Cryopreservation, although very useful from a production and conservation point of view, produces several types of damage to germ cells, especially to spermatozoa. Therefore, during the last decades huge efforts were done trying to reduce cryodamage, identify its causes and consequences to the sperm cell.

There are several constituents in a germ cell that can be analyzed. These analyses depend on the objective, possible applications, and on the methodology and available equipment. They can be from a simple motility analysis using a subjective scoring (normally used in fish farms) to more sophisticated analysis of motility, involving specific software (e.g., CASA system) or image analysis systems (Image J) [39]. Sperm quality can also be assessed by its constituents: seminal plasma and spermatozoa, indicating damage in specific cellular structures or in the entire cell, by releasing its constituents in the seminal plasma [39]. Determination of seminal plasma constituents and its variations such as lytic, oxidative, metabolic and apoptotic enzymes, metabolites, sugars, vitamins, amino acids, fatty acids, and other inorganic compounds can provide very useful information on sperm status [39]. Other characteristics of these cells should also be evaluated in order to assess sperm fertilization ability, especially when this assay is difficult to perform due to egg availability or long-term embryonic development. Most tests describing cell viability or mitochondrial status are currently performed with the use of fluorescent probes combined with microscopy or flow cytometry [40]. There is a huge list, depending on the main objective and available supplies and equipment, and can be from the determination of reactive oxygen species (ROS) levels, MDA concentration, detecting oxidative events at protein or lipid level, or simple cell structure impairments or ruptures. The assessment of sperm DNA is another test that should be performed since will provide information on sperm fitness being important for offspring quality. Methods to evaluate chromatin integrity include the comet assay (single cell gel electrophoresis), TUNEL (terminal deoxynucleotidyl transferase-nick-end-labeling), SCSA (sperm chromatin structure assay), and the analysis of specific DNA sequences using qPCR [41–43]. In recent years, and as an attempt to define bases for explaining other sperm quality markers at different levels, the characterization and presence of certain transcripts has been used [43]. These new tests can reveal the true meaning of spermatozoa quality.

All these techniques, although very useful in the evaluation of sperm quality and offspring viability, are still on a laboratory scale and need to be adapted for industry. Efforts are still required to transfer current technologies and make sperm banking accessible for fish producers [44]. In the last years, some national and international networks have been trying to get close contact between researchers and fish farm industries [45]. Although, there are still a lot of work to do, this topic should be considering as a priority step. It is also clear that although fundamental research is moving forward in the design of new methods and techniques to characterize sperm quality, it is important to focus on a process of scaling-up technologies to encourage companies and fish farmers in their use. Aquaculture needs to benefit from reproductive biotechnology.

7.2. Quality evaluation in sperm from Brazilian species

According to Viveiros, Orfão [8] there are 18 Brazilian native species with high potential for cryopreservation, and indeed, several protocols have been developed and the postthaw quality assessed in most of those species (**Table 1**). Most of the work has been conducted in three main species, *P. lineatus, Brycon orbignyanus*, and *C. macropomum*. The first two more dedicated to conservation and slightly less to production and the *C. macropomum* more dedicated to production. In terms of postthaw quality analysis, research conducted in the last years, especially in these species gained high relevance and allowed to detect specific damage associated with cryopreservation. A subjective analysis of motility is still the main parameter analyzed, which in some cases can be responsible for the high variability of results obtained by different groups working in the same species [44]. The complementary analysis of other parameters such as cell viability using flow cytometer of fluorescent microscopy can help in this matter. Recent work in the previous mentioned species allowed to distinguish methyl glycol as the best cryoprotectant for *P. lineatus* and *B. orbignyanus* in terms of motility (63 and 72% motile cells, respectively) and cell viability (57 and 68% viable cells, respectively) [46]. A similar analysis concluded that DMSO and DMF (5 and 8%, respectively) were the best choices for freezing *C. macropomum* sperm, yielding motility and viability rates higher than 50%. Although, some damage was detected in postthaw samples, fertility, and hatching rates were not compromised (91.6 and 87.6%) [24, 47]. Therefore, it seems clear that although it is very important to perform a complete analysis of sperm using several indicators of quality such as motility, cell viability, and functionality and DNA status, fertility, and especially hatching rates are the ultimate tests to characterize the success of any cryopreservation protocol. These assays should be performed in order to demonstrate the viability of this technology to fish farmers, showing its applicability in aquaculture industry.

Species	Assessed parameter	Level of knowledge[1]	References
Brycon amazonicus	Motility (Subjective)	Low	Casallas, Robles [56]
	Motility (Subjective)		Cruz-Casallas, Medina-Robles [57]
	Motility (Subjective), fertilization rate		Velasco-Santamaría, Medina-Robles [27]
Brycon insignis	Motility (Subjective-CASA)	Low	Viveiros, Orfão [58]
	Motility (Subjective), viability		Viveiros, Amaral [59]

Species	Assessed parameter	Level of knowledge[1]	References
Brycon nattereri	Motility (Subjective), ultrastructure	Low	Viveiros, Maria [19]
	motility (Subjective)		Oliveira, Viveiros [60]
Brycon opalinus	Motility (Subjective), viability	Low	Viveiros, Orfão [58]
Brycon orbignyanus	Motility (CASA)	Medium	López, Leal [61]
	Motility (CASA)		Viveiros, Gonçalves [62]
	Motility (CASA), viability, mitochondrial functionality		Viveiros, Nascimento [46]
	Motility, morphology (Subjective)		Andrade, de Jesus [63]
	Motility, morphology (Subjective)		Galo, Streit Jr [64]
	motility (Subjective)		Viveiros, Maria [17]
	Motility (Subjective), fertilization rate		Maria, Viveiros [16]
	Motility (Subjective)		Carolsfeld, et al. [7]
Brycon orthoenia	Motility (Subjective)	Low	Melo and Godinho [65]
Colossoma macropomum	Motility (Subjective), fertilization rate, DNA methylation	High	de Mello, Garcia [66]
	motility (CASA)		Melo-Maciel, Leite-Castro [67]
	Motility (Subjective), fertilization rate, viability, mitochondrial functionality, DNA integrity		Garcia, Vasconcelos [20]
	Motility (Subjective), fertilization rate, viability, mitochondrial functionality, DNA integrity		Varela Junior, Goularte [24]
	Motility (Subjective), fertilization rate, viability, mitochondrial functionality, DNA integrity		Varela Junior, Corcini [47]
	motility (CASA)		Leite, Oliveira [21]
	Motility (Subjective)		Menezes, Queiroz [68]
Leporinus macrocephalus	Motility (Subjective)	Low	Ribeiro and Godinho [69]
Leporinus obtusidens	Motility (Subjective)	Low	Viveiros, Maria [17]
	Motility (Subjective), fertilization rate		Taitson, Chami [26]
Piaractus brachypomus	Motility (Subjective-CASA)	Low	Nascimento, Maria [70]

Species	Assessed parameter	Level of knowledge[1]	References
Piaractus mesopotamicus	Motility, morphology (Subjective)	Low	Andrade, de Jesus [63]
	Morphology (Subjective)		Paulino, Murgas [71]
	Motility (Subjective), morphology (Subjective)		Streit Jr, Benites [72]
	Motility (Subjective)		Carolsfeld, et al. [7]
Prochilodus lineatus	Motility (CASA)	Medium	Viveiros, Gonçalves [62]
	Motility (CASA), viability, mitochondrial functionality		Viveiros, Nascimento [46]
	Motility (Subjective), morphology (Subjective)		Vasconcelos, Felizardo [73]
	Motility, morphology (Subjective)		Andrade, de Jesus [63]
	Motility (Subjective)		Navarro, Navarro [74]
	Motility (Subjective)		Paula, Andrade [75]
	Morphology (Subjective), fertilization rate		Miliorini, Murgas [76]
	Motility (Subjective), morphology (Subjective)		Felizardo, Mello [77]
	Motility (CASA), fertilization rate		Viveiros, Nascimento [15]
	Motility (Subjective), fertilization rate		Viveiros, Orfão [78]
	Motility (Subjective)		Viveiros, Maria [17]
	Motility (Subjective)		Murgas, Miliorini [79]
	Motility (Subjective)		Carolsfeld, et al. [7]
Pseudoplatystoma corruscans	Motility (Subjective), fertilization rate	Low	Cruz-Casallas, Medina-Robles [57]
Rhamdia quelen	Motility (CASA), fertilization rate	Low	Adames, de Toledo [23]
Salminus brasiliensis	Motility (Subjective), fertilization rate	Low	Viveiros, Oliveira [18]
	Fertilization rate		Zanandrea, Weingartner [80]
	Motility (Subjective)		Zanandrea, Weingartner [81]

[1] Level of knowledge according to the used methodology on the postthaw sperm analysis of each species. Data have been collected from Web of Science platform since papers published from 2000 onwards by using as key words "species name + cryopreservation."

Table 1. Methods used for postthaw evaluation of sperm quality in freshwater Brazilian species.

8. Difficulties in broodstock standardization and acquisition of wild or genetic improved fish

Standardizing a broodstock requires much more than just buying animals from different locations. To address this issue there is a need for a holistic knowledge of the whole process. The technical efficiency of a germplasm bank goes far beyond the knowledge of physicochemical characteristics of semen or a well-performed cryopreservation protocol. The first thing to be defined is the goal of germplasm bank: for restocking or fingerlings production to fish farming. When the goal is to restock environments, the most important aspect is the bank composition by animals that have the greatest possible genetic diversity. We should include here wild animals, even though they do not have a good seminal quality, but that make up the nearest setting of natural environment. On the other hand, when we set up a broodstock for commercial production of fingerlings we should seek animals with a greater genetic standardization, once the results will also have less variation. Thus, to establishment of a commercial broodstock we should analyze the genetic similarity degree of the animals before selecting them.

Insert wild fish in a broodstock whose objective is to produce fingerlings for fish farming can be a terrible mistake. Wild animals can carry parasites and therefore introduce them in the farm broodstock. Another important factor that should be considered is the broodstock domestication. According to Ruzzante [48], domestication is a process of adaptation of organisms in the man-made environment. We had an interesting experience a few years ago, when we use a semen bank from wild *P. mesopotamicus* to produce fingerlings from eighth-generation captive females. The generated fingerlings were darker and presented a different behavior compared to those animals produced from domesticated parents. It is important to note that this was just an observation and not an experiment.

The formation of a germplasm bank should be conducted with the support of molecular techniques, which are still timidly used in Brazil [49]. Although being observed a high degree of genetic variability in broodstock of native species that have been formed in the country, their future efficiency may be compromised. We can cite as example the *C. macropomum* broodstock from two large farms in northern Brazil, where there was no problem of genetic variability, even with both having the same founder effect [50]. In another example, Jacometo et al. [51] reported that in addition to the high genetic variability in *C. macropomum* broodstock it was also observed a moderate differentiation and low genetic distance within themselves. The broodstock formation in this species occurred in 1972 [52].

The broodstock of *B. orbignyanus* [53] and *P. mesopotamicus* [54] used for restocking showed no genetic variability problems. We emphasize that even not having a compromising genetic variability degree it is important to consider the degree of genetic distance of the animals to be used as semen donors; once by accident the collection of semen from sibling animals can occur, which could be compromising depending on the ultimate goal of the germplasm bank.

This context serves to show how much we should be careful in setting up a germplasm bank from a random broodstock. It must be considered from the selected animals: the purpose of the genebank, age, genetic origin, and location; consanguinity degree, etc. A good application example of these concepts can be found in the article published by Streit et al. [55], in which two germplasm banks were created for use in breeding programs of two Brazilian fish species, *C. macropomum* and *Pseudoplatystoma reticulatum*.

9. The challenge of making viable the use of cryopreserved semen in Brazil

The first question to be asked is: Are there any germplasm banks in Brazil? Yes, there are! They began to be formed in the 1980s by Professor Hugo Godinho and the visit of two Canadian researchers, Brian Harvey and Joachim Carolsfeld, who started extensive work with Brazilian researchers. Among their actions, they have developed cryoprotocols for some species, already recorded in this chapter [7]. For over 30 years some banks of native species semen were formed and has been used mainly as deposit of genetic material for research at universities laboratories and other ones as genetic reserve for restocking (personal communication from: Bombardelli, R.; Maria, A.N.; Murgas, L.D.; Ninhaus-Silveira, A.; Resende, E.K.; Ribeiro, R.P.; Streit Jr., D.P., and Viveiros, A.).

A relevant question concerns the active use of germplasm banks in the country, either for commercial purposes (fish farming) or conservation. Popularize the use of cryopreserved fish semen in the country as far as the use of bovine or equine semen, for instance, is still far away. Some conditions are crucial to this fact: the culture of fingerlings producers not to use a germplasm bank; have actually consistent cryopreservation protocols that can be used effectively; unfamiliarity of the market to absorb the product (frozen semen) and especially the real need for the use of cryopreserved semen in fingerling production farms.

Almost all Brazilian farms are unaware of the genetic characteristics of their broodstocks. Thus, there would be no sense to use semen from another location without any objective criteria to explore genetic gain. Moreover, it is significant that most of the Brazilian migratory species present reproductive asynchrony in captivity. It is quite common to note that at the end of reproductive season there is availability of mature females but not males. Thereby, the germplasm bank could meet this need. However, what is the cost to maintain a germplasm bank? There would be any market demand? Are there concluding protocols ready, tested, and efficient to be used for the main migratory species that are farmed? The last question is not that simple to answer. The quality parameters from cryopreserved semen still vary widely. From this reflection, we must establish priority in research in order to meet the demands, even if they are as wide as the number of species that make up our ichthyofauna. Focus research efforts to establish a secure protocol for commercial application purposes should be a goal.

The germplasm bank significantly facilitates the establishment of a breeding program for fish [55]. Surely, the semen distribution from known strain animals will contribute to spread of genebanks and there may be a specific market to the product cryopreserved semen. We believe that in the future the industry will be aged enough to realize it, especially those fingerlings buyers that will seek fish with known pedigree, even if it is necessary to pay a little more for them. Soon, the company that offers high-quality genetic material can explore this market.

Regarding germplasm banks with restocking purposes, most are circumstantial. They have been created through research and development (R&D) projects, largely in partnership with electric power companies. These gene banks are maintained while the project is running and later are reduced to a small number of samples (as a collection), especially due to maintenance costs. The genebanks that are intended for restocking, nowadays mostly consist of semen from species that are also farmed. On the other hand, it would need to extend to other species, such as the large Amazon catfishes: *Phractocephalus hemioliopterus; Brachyplathystoma filamentosum* and *B. Flavicans,* and others as *Zungaro jahu, Surubim lima* and *Hemisorubim platyrhynchos*; in addition to the species endangered by numerous factors such as the construction of hydro-electric dams, overfishing, and pollution of the aquatic environment.

10. Conclusions, perspectives and future challenges

Although, cryopreservation of Brazilian species has been improving in the last years with more publications coming to this field of research and especially with a far advanced analysis on cryodamage, there is still a long way to follow. This is truly mainly in those species where commercial interests and aquaculture benefits can arise or in species where conservation is a main challenge. One of the main objectives to pursue is for sure a more exhaustive analysis of sperm quality, adapting techniques that are already used in other species. This will give guarantees of quality for producers that will set out the best protocols defined. Another point to take in consideration is standardization of procedures and reporting, not only related with cryopreservation protocol, but especially with the analysis performed. This lack in standardization has been already identified in some of the most reported species around the world and over the last years it has been a struggle to joint researchers in following certain rules. Building a database of standardized information for semen analysis would be crucial. What has been discussed throughout this chapter has significant relevance in the compression of events that concern fish semen cryopreservation in Brazil. The advances made in recent years by efforts of research groups are commendable. However, there still a lot to do, and joint efforts through networking activities can be a good opportunity, not only for training young researchers and technicians, but also to find solutions to the main problems.

Acknowledgements

The authors would like to thank COST action AQUAGAMETE (FA1205) Food and Agriculture and BAESA/ENERCAN power companies. Special thanks should be given to Dr. Alexandre N. Maria, Dr. Alexandre Ninhaus-Silveira, Dr. Ana Viveiros, Dr. Emiko K. Resende, Dr. Luis D. Murgas, Dr. Ricardo P. Ribeiro, and Dr. Robie Bombardelli for providing information that significantly contributed to this chapter.

Author details

Raycon R.F. Garcia[1,2*], Danilo P. Streit Jr[1], Elsa Cabrita[2] and Leandro C. Godoy[1]

*Address all correspondence to: raycongarcia@live.com

1 Graduate Program in Animal Science, Federal University of Rio Grande do Sul, Porto Alegre, Brazil

2 CCMAR, University of Algarve, Faro, Portugal

References

[1] Coser AM, Godinho H, Ribeiro D. Cryogenic preservation of spermatozoa from *Prochilodus scrofa* and *Salminus maxillosus*. Aquaculture. 1984;37(4):387–390.

[2] Fogli da Silveira W, Kavamoto ET, Narahara MY. Avaliação da qualidade e criopreservação em forma de "pellets" do sêmen do bagre, *Rhamdia hilarii* (Valenciennes, 1840). Boletim do Instituto de Pesca. 1985;12(4):7–11.

[3] Coser AML, Godinho HP, Torquato VC. Criopreservaçao do sêmen do peixe piau *Leporinus silvestrii* (Boulanger, 1902). Arquivo Brasileiro de Medicina Veterinária e Zootecnia. 1987;39:37–42.

[4] Carolsfeld J, Harvey B, Silveira W, Kavamoto E, Ramos S, Silveira A. Criopreservação do semen de pacu, *Piaractus mesopotamicus* Holmberg, 1987. Boletim Técnico do CEPTA. 1990;3:1–4.

[5] Fogli da Silveira W, Kavamoto E, Cestarolli M, Godinho H, Ramos S, Silveira A. Avaliação espermática, preservação criogênica e fertilidade do sêmen do pacu, *Piaractus mesopotamicus* (Holmberg, 1887), proveniente de reprodução induzida. Boletim do Instituto de Pesca. 1990;17:1–3.

[6] Farias JO, Nunes JF, de Carvalho MAM, de Mello Salgueiro CC. In vitro and in vivo evaluation of the Tambaqui semen "*Colossoma macropomum*" conserved at environmental temperature and cryo-preserved in coconut water. Revista Científica de Produção Animal. 1999;1(1):44–58.

[7] Carolsfeld J, Godinho HP, Zaniboni Filho E, Harvey BJ. Cryopreservation of sperm in Brazilian migratory fish conservation. Journal of Fish Biology. 2003;63(2):472–489.

[8] Viveiros ATM, Orfão LH, Leal M. Biologia e Conservação de Espermatozoides. In: Baldisserotto B, Cyrino J, Urbinati EC, editors. Biologia e Fisiologia de Peixes Neotropicais de Água Doce. Jaboticabal: FUNEP/UNESP; 2014. pp. 307–327.

[9] Lahnsteiner F, Mansour N, Caberlotto S. Composition and metabolism of carbohydrates and lipids in *Sparus aurata* semen and its relation to viability expressed as sperm

motility when activated. Comparative Biochemistry and Physiology Part B: Biochemistry and Molecular Biology. 2010;157(1):39–45.

[10] Bergeron A, Vandenberg G, Proulx D, Bailey JL. Comparison of extenders, dilution ratios and theophylline addition on the function of cryopreserved walleye semen. Theriogenology. 2002;57(3):1061–1071.

[11] Watson P. The roles of lipid and protein in the protection of ram spermatozoa at 5 degrees C by egg-yolk lipoprotein. Journal of Reproduction and Fertility. 1981;62(2): 483–492.

[12] Bergeron A, Manjunath P. New insights towards understanding the mechanisms of sperm protection by egg yolk and milk. Molecular Reproduction and Development. 2006;73(10):1338–1344.

[13] Carneiro PC, Azevedo HC, Santos JP, Maria AN. Cryopreservation of tambaqui (*Colossoma macropomum*) semen: extenders, cryoprotectants, dilution ratios and freezing methods. CryoLetters. 2012;33(5):385–393.

[14] Orfão LH, Maria AN, Nascimento AF, Isaú ZA, Viveiros A. Sperm fertility of the subtropical freshwater streaked prochilod *Prochilodus lineatus* (Characiformes) improved after dilution and cold storage. Aquaculture Research. 2010;41(10):679–687.

[15] Viveiros A, Nascimento A, Orfão L, Isaú Z. Motility and fertility of the subtropical freshwater fish streaked prochilod (*Prochilodus lineatus*) sperm cryopreserved in powdered coconut water. Theriogenology. 2010;74(4):551–556.

[16] Maria A, Viveiros A, Freitas R, Oliveira A. Extenders and cryoprotectants for cooling and freezing of piracanjuba (*Brycon orbignyanus*) semen, an endangered Brazilian teleost fish. Aquaculture. 2006;260(1):298–306.

[17] Viveiros A, Maria AN, Orfao LH, Carvalho MAM, Nunes JF. Powder coconut water (ACP®) as extender for semen cryopreservation of Brazilian migratory fish species. Cybium. 2008;32(2):215.

[18] Viveiros A, Oliveira A, Maria A, Orfão L, Souza J. Sensibilidade dos espermatozoides de dourado (*Salminus brasiliensis*) a diferentes soluções crioprotetoras. Arquivo Brasileiro de Medicina Veterinária e Zootecnia. 2009;61(4):883–889.

[19] Viveiros ATM, Maria AN, Amaral TB, Orfão LH, Isau ZA, Veríssimo-Silveira R. Spermatozoon ultrastructure and sperm cryopreservation of the Brazilian dry season spawner fish pirapitinga, *Brycon nattereri*. Aquaculture Research. 2012;43(4): 546–555.

[20] Garcia RRF, Vasconcelos ACN, Povh JA, Oberst ER, Varela Jr AS, Corcini CD, et al. Functional integrity of *Colossoma macropomum* (Cuvier, 1816) sperm cryopreserved with enriched extender solutions. Neotropical Ichthyology. 2015;13(3): 599–606.

[21] Leite LV, Oliveira F, Nunes LT, Nunes JF, Salmito-Vanderley CSB. Sperm cryopreservation of tambaqui in powdered coconut water (ACP™) added of egg yolk. Revista Brasileira de Engenharia de Pesca. 2011;6:23–29.

[22] Mazur P. Principles of Cryobiology. In: Fuller BJ, Lane N, Benson EE, editors. Life in the Frozen State. Boca Raton, FL: CRC Press; 2004. pp. 3–65.

[23] Adames MS, de Toledo CPR, Neumann G, Buzzi AH, Buratto CN, Piana PA, et al. Optimization of the sperm: oocyte ratio and sperm economy in the artificial reproduction of *Rhamdia quelen* using fructose as a sperm motility modulator. Animal Reproduction Science. 2015;161:119–128.

[24] Varela Junior A, Goularte K, Alves J, Pereira F, Silva E, Cardoso T, et al. Methods of cryopreservation of tambaqui semen, *Colossoma macropomum*. Animal Reproduction Science. 2015;157:71–77.

[25] Maria AN, Carvalho ACM, Araújo RV, Santos JP, Carneiro PCF, Azevedo HC. Use of cryotubes for the cryopreservation of tambaqui fish semen (*Colossoma macropomum*). Cryobiology. 2015;70(2):109–114.

[26] Taitson P, Chami E, Godinho H. Gene banking of the neotropical fish *Leporinus obtusidens* (Valenciennes, 1836): a protocol to freeze its sperm in the field. Animal reproduction science. 2008;105(3):283–291.

[27] Velasco-Santamaría YM, Medina-Robles VM, Cruz-Casallas PE. Cryopreservation of yamú (*Brycon amazonicus*) sperm for large scale fertilization. Aquaculture. 2006;256(1): 264–271.

[28] Cabrita E, Robles V, Alvarez R, Herráez M. Cryopreservation of rainbow trout sperm in large volume straws: application to large scale fertilization. Aquaculture. 2001;201(3): 301–314.

[29] Lahnsteiner F, Weismann T, Patzner R. Methanol as cryoprotectant and the suitability of 1.2 ml and 5 ml straws for cryopreservation of semen from salmonid fishes. Aquaculture Research. 1997;28(6):471–479.

[30] Ingermann R, Schultz C, Kanuga M, Wilson-Leedy J. Metabolism of motile zebrafish sperm. Comparative Biochemistry and Physiology Part A: Molecular and Integrative Physiology. 2011;158(4):461–467.

[31] Rurangwa E, Kime D, Ollevier F, Nash J. The measurement of sperm motility and factors affecting sperm quality in cultured fish. Aquaculture. 2004;234(1):1–28.

[32] Alavi S, Cosson J. Sperm motility in fishes. (II) Effects of ions and osmolality: a review. Cell biology international. 2006;30(1):1–14.

[33] Cosson J. The ionic and osmotic factors controlling motility of fish spermatozoa. Aquaculture International. 2004;12(1):69–85.

[34] Alavi SH, Rodina M, Viveiros A, Cosson J, Gela D, Boryshpolets S, et al. Effects of osmolality on sperm morphology, motility and flagellar wave parameters in Northern pike (*Esox lucius L.*). Theriogenology. 2009;72(1):32–43.

[35] Hu J, Zhang Y, Zhou R. Changes in extracellular osmolality initiate sperm motility in freshwater teleost rosy barb *Puntius conchonius*. Theriogenology. 2009;72(5):704–710.

[36] Bombardelli RA, Mörschbächer EF, Campagnolo R, Sanches EA, Syperreck MA. Insemination dose for artificial fertilization of grey jundiá oocytes, *Rhamdia* quelen (Quoy & Gaimardm, 1824). Revista Brasileira de Zootecnia. 2006;35(4):1251–1257.

[37] Leite L, Melo M, Oliveira F, Pinheiro J, Campello C, Nunes J, et al. Determination of insemination dose and embryonic development in the artificial fertilization of tambaqui (*Colossoma macropomum*). Arquivo Brasileiro de Medicina Veterinária e Zootecnia. 2013;65(2):421–429.

[38] Sanches EA, Bombardelli RA, Baggio DM, Souza BEd. Insemination dose for artificial fertilization of dourado oocytes. Revista Brasileira de Zootecnia. 2009;38(11):2091–2098.

[39] Cabrita E, Robles V, Herráez P. Methods in Reproductive Aquaculture: Marine and Freshwater Species. Biology Series: CRC Press (Taylor & Francis Group); Boca Raton, FL. 2009. 574 pp.

[40] Migaud H, Bell G, Cabrita E, McAndrew B, Davie A, Bobe J, et al. Broodstock management and gamete quality in temperate fish. Reviews in Aquaculture. 2013;5(1):194–223.

[41] Cabrita E, Martínez-Páramo S, Gavaia PJ, Riesco M, Valcarce D, Sarasquete C, et al. Factors enhancing fish sperm quality and emerging tools for sperm analysis. Aquaculture. 2014;432:389–401.

[42] Cabrita E, Sarasquete C, Martínez-Páramo S, Robles V, Beirao J, Pérez-Cerezales S, et al. Cryopreservation of fish sperm: applications and perspectives. Journal of Applied Ichthyology. 2010;26(5):623–35.

[43] Robles V, Herráez M, Labbé C, Cabrita E, Psenicka M, Valcarce D, et al. Molecular basis of sperm quality. General and Comparative Endocrinology. 2016;(*in press*)(doi.org/10.1016/j.ygcen.2016.04.026).

[44] Martínez-Páramo S, Horváth Á, Labbé C, Zhang T, Robles V, Herráez M, et al. Cryobanking of aquatic species. Aquaculture. 2016; (*in press*) (doi:10.1016/j.aquaculture.2016.05.042)

[45] Asturiano JF, Cabrita E, Horváth Á. Progress, challenges and perspectives on fish gamete cryopreservation: a mini review. General and Comparative Endocrinology. 2016. (doi:10.1016/j.ygcen.2016.06.019)

[46] Viveiros AT, Nascimento AF, Leal MC, Gonçalves AC, Orfão LH, Cosson J. Methyl glycol, methanol and DMSO effects on post-thaw motility, velocities, membrane

integrity and mitochondrial function of *Brycon orbignyanus* and *Prochilodus lineatus* (Characiformes) sperm. Fish Physiology and Biochemistry. 2015;41(1):193–201.

[47] Varela Junior A, Corcini C, Gheller S, Jardim R, Lucia T, Streit Jr D, et al. Use of amides as cryoprotectants in extenders for frozen sperm of tambaqui, *Colossoma macropomum*. Theriogenology. 2012;78(2):244–251.

[48] Ruzzante DE. Domestication effects on aggressive and schooling behavior in fish. Aquaculture. 1994;120(1):1–24.

[49] Hilsdorf AWS, Orfão LH. General aspects of genetic improvement of fish in Brazil. Revista Brasileira de Zootecnia. 2011;40:317–324.

[50] Lopes TdS, Streit Jr DP, Ribeiro RP, Povh JA, Lopera Barrero NM, Vargas LDM, et al. Genetic diversity of *Colossoma macropomum* broodstock. Arquivo Brasileiro de Medicina Veterinária e Zootecnia. 2009;61(3):728–735.

[51] Jacometo CB, Lopera Barrero NM, Rodriguez-Rodriguez MDP, Gomes PC, Povh JA, Streit Júnior DP, et al. Genetic variability of tambaqui (Teleostei: Characidae) from different regions of Brazil. Pesquisa Agropecuaria Brasileira. 2010;45(5):481–487.

[52] Calcagnotto D, Toledo-Filho SdA. Loss of genetic variability at the transferrin locus in five hatchery stocks of tambaqui (*Colossoma macropomum*). Genetics and Molecular Biology. 2000;23(1):127–130.

[53] Lopera-Barrero N, Vargas L, Sirol R, Ribeiro R, Povh J, Mangolin C. Genetic characterization of Brycon orbignyanus using the semi-natural system. Arquivo Brasileiro de Medicina Veterinária e Zootecnia. 2010;62(1):184–191.

[54] Povh J, Ribeiro R, Lopera-Barrero N, Gomes P, Blanck D, Vargas L, et al. Monitoring of the genetic variability of pacu, *Piaractus mesopotamicus*, of the stock enhancement program of the Paranapanema River. Arquivo Brasileiro de Medicina Veterinária e Zootecnia. 2009;61(5):1191–1195.

[55] Streit Jr DP, Fornari D, Povh J, Godoy L, de Mello F, Oliveira C, et al. Germplasm banking and its role in the development of the fish genetic improvement programme in Brazil. CryoLetters. 2015;36(6):399–404.

[56] Casallas PEC, Robles VMM, Santamaría YV. Evaluation of different cryoprotectors for cryopreservation of yamú (*Brycon amazonicus*) spermatozoa. Revista Colombiana de Ciencias Pecuarias. 2006;19(2):152–159.

[57] Cruz-Casallas PE, Medina-Robles VM, Velasco-Santamaría YM. Protocol for cryopreservation of yamú: *Brycon amazonicus*. Revista Colombiana de Ciencias Pecuarias. 2006;19(2):146–151.

[58] Viveiros A, Orfão L, Nascimento A, Corrêa F, Caneppele D. Effects of extenders, cryoprotectants and freezing methods on sperm quality of the threatened Brazilian

freshwater fish pirapitinga-do-sul *Brycon opalinus* (Characiformes). Theriogenology. 2012;78(2):361–368.

[59] Viveiros AT, Amaral TB, Orfão LH, Isau ZA, Caneppele D, Leal MC. Sperm cryopreservation of tiete tetra *Brycon insignis* (Characiformes): effects of cryoprotectants, extenders, thawing temperatures and activating agents on motility features. Aquaculture Research. 2011;42(6):858–865.

[60] Oliveira A, Viveiros A, Maria A, Freitas R, Izaú Z. Sucess of cooling and freezing of pirapitinga (*Brycon nattereri*) semen. Arquivo Brasileiro de Medicina Veterinária e Zootecnia. 2007;59(6):1509–1515.

[61] López D, Leal M, Viveiros A. Extender composition and osmolality affects post-thaw motility and velocities of piracanjuba *Brycon orbignyanus* (Valenciennes, 1850)(Characiformes) sperm. Journal of Applied Ichthyology. 2015;31(S1):114–118.

[62] Viveiros AT, Gonçalves A, Nascimento AF, Leal MC. Fresh, equilibrated and post-thaw sperm quality of *Brycon orbignyanus* (Valenciennes, 1850) and *Prochilodus lineatus* (Valenciennes, 1837) treated with either salmon GnRHa and domperidone or pituitary extract. Neotropical Ichthyology. 2015;13(1):157–164.

[63] Andrade EdS, de Jesus P, Aparecida D, Felizardo VdO, Murgas LDS, Veras GC. Milt cryopreservation for rheophilic fish threatened by extinction in the Rio Grande, Brazil. CryoLetters. 2014;35(1):8–14.

[64] Galo JM, Streit Jr DP, Sirol RN, Ribeiro RP, Digmayer M, Andrade V, et al. Spermatic abnormalities of piracanjuba *Brycon orbignyanus* (Valenciennes, 1849) after cryopreservation. Brazilian Journal of Biology. 2011;71(3):693–699.

[65] Melo F, Godinho H. A protocol for cryopreservation of spermatozoa of the fish *Brycon orthotaenia*. Animal Reproduction. 2006;3(3):380–385.

[66] de Mello F, Garcia JS, Godoy LC, Depincé A, Labbé C, Streit Jr DP. The effect of cryoprotectant agents on DNA methylation patterns and progeny development in the spermatozoa of *Colossoma macropomum*. General and Comparative Endocrinology. 2016; in press.

[67] Melo-Maciel M, Leite-Castro L, Leite J, Oliveira M, Almeida-Monteiro P, Nunes J, et al. Aloe vera in the cryopreservation of tambaqui (*Colossoma macropomum*) sperm. Arquivo Brasileiro de Medicina Veterinária e Zootecnia. 2015;67(3):945–949.

[68] Menezes JTB, Queiroz L, Doria CRdC, Menezes Jr J. Sperm evaluation of tambaqui, *Colossoma macropomum* (Cuvier, 1818), after thawing. Acta Amazonica. 2008;38(2): 365–368.

[69] Ribeiro R, Godinho H. Testicular sperm cryopreservation of the teleost'piau-açu'*Leporinus macrocephalus*. Arquivo Brasileiro de Medicina Veterinária e Zootecnia. 2003;55(1):75–79.

[70] Nascimento A, Maria A, Pessoa N, Carvalho M, Viveiros A. Out-of-season sperm cryopreserved in different media of the Amazonian freshwater fish pirapitinga (*Piaractus brachypomus*). Animal Reproduction Science. 2010;118(2):324–329.

[71] Paulino M, Murgas L, Felizardo V, Freitas R. Abnormalities of sperm after thawing *Piaractus mesopotamicus* using different methodologies. Arquivo Brasileiro de Medicina Veterinária e Zootecnia. 2012;64(6):1591–1596.

[72] Streit Jr DP, Benites C, de Moraes GV, Ribeiro RP, Sakaguti ES, Caldieri RF. Semen of pacu (*Piaractus mesopotamicus*) criopreserved used diluent for swine semen. Ciência Animal Brasileira. 2006;7(3):289–297.

[73] Vasconcelos ACN, Felizardo VdO, Ferreira A, de Carvalho S, Garcia RRF, Ramos SE, et al. Cryopreservation of *Prochilodus lineatus* semen: effect of cryoprotectants combination. Boletim do Instituto de Pesca. 2015;41(*esp*.):817–824.

[74] Navarro RD, Navarro FKSP, de Oliveira Felizardo V, Murgas LDS, de Souza Andrade E. Semen quality of curimba (*Prochilodus lineatus*) cryopreserved with vitamins. Acta Scientiarum Technology. 2014;36(1):55–60.

[75] Paula DA, Andrade ES, Murgas LD, Felizardo VO, Winkaler EU, Zeviani W, et al. Vitamin E and reduced glutathione in *Prochilodus lineatus* (curimba) semen cryopreservation (Characiformes: Prochilodontidae). Neotropical Ichthyology. 2012;10(3):661–665.

[76] Miliorini AB, Murgas LDS, Rosa PV, Oberlender G, Pereira GJM, da Costa DV. A morphological classification proposal for curimba (*Prochilodus lineatus*) sperm damages after cryopreservation. Aquaculture Research. 2011;42(2):177–187.

[77] Felizardo V, Mello R, Murgas L, Andrade E, Drumond M, Rosa P. Effect of cryopreservant combinations on the motility and morphology of curimba (*Prochilodus lineatus*) sperm. Animal Reproduction Science. 2010;122(3):259–263.

[78] Viveiros A, Orfão L, Maria A, Allaman I. A simple, inexpensive and successful freezing method for curimba *Prochilodus lineatus* (Characiformes) semen. Animal Reproduction Science. 2009;112(3):293–300.

[79] Murgas LDS, Miliorini AB, Freitas R, Pereira GJM. Cryopreservation of curimba (*Prochilodus lineatus*) semen after addition of different diluters, activators and cryoprotectants. Revista Brasileira de Zootecnia. 2007;36(3):526–531.

[80] Zanandrea ACV, Weingartner M, Zaniboni-Filho E. Induced reproduction of dourado (*Salminus brasiliensis*): fertilization with sperm cryopreserved in ACP-104. Acta Scientiarum Animal Sciences. 2016;38(1):17–22.

[81] Zanandrea CV, Weingartner M, Zaniboni-Filho E. Use of ACP-104 at different dilutions for sperm cryopreservation of dourado, *Salminus brasiliensis*. Journal of the World Aquaculture Society. 2014;45(1):82–87.

Chapter 5

Cryopreservation of Nile Tilapia (*Oreochromis niloticus*) Sperm

Yusuf Bozkurt and İlker Yavaş

Additional information is available at the end of the chapter

http://dx.doi.org/10.5772/64835

Abstract

The main aim of this study is to determine the effect of the straw volume (0.25 vs. 0.5 mL) on Nile tilapia sperm quality after cryopreservation. Sperm was frozen according to conventional slow freezing procedure and diluted at ratio of 1:3 with ionic extender containing 350 mM glucose and 30 mM Tris containing 10% dimethylacetamide. Diluted semen was equilibrated at 4°C for 10 min and drawn into 0.25-mL or 0.5-mL plastic straws and sealed with polyvinyl alcohol. Samples were frozen 3 cm above of the liquid nitrogen surface and exposed to the liquid nitrogen vapor (≈−140°C) for 10 min. After this, frozen sperm cells were kept into the liquid nitrogen container (−196°C). The frozen sperm in different volume of straws were thawed in a water bath at 30°C for 20 s (0.25-mL straws) or at 30°C for 30 s (0.5-mL straws), respectively. Fertilization was conducted using 1×10^5 spermatozoa/egg ratio with each straw type. The findings of the present study indicated that cryopreservation of sperm in glucose-Tris–based extender using 0.5-mL straws improved post-thaw progressive motility, duration of progressive motility, and fertilization results ($P<0.01$). On the other hand, differences in term of post-thaw cell viability was not significant among the treatments ($P>0.01$). In conclusion, our results suggest that Nile tilapia sperm can be successfully cryopreserved in Tris-based extenders supplemented with glucose containing 10% dimethylacetamide in 0.5-mL straws.

Keywords: *Oreochromis niloticus*, sperm, cryopreservation, straw volume, dimethylacetamide

1. Introduction

Cryopreservation biotechnology has important roles for aquaculture industry and also for conservation of aquatic genetic resources. In this field, sperm cryopreservation has been used for transporting of genetic material between facilities, optimal using of gametes in aquaculture, reducing risk of spreading infections, performing of hybridization studies, conserving of protecting endangered species, and also for conserving of biodiversity [1, 2].

Cryopreservation technique involves addition of cryoprotectants to the extender and freezing and thawing of sperm samples, which may result in some damage to the spermatozoa and may decrease egg fertilization rate. Therefore, before cryopreservation of spermatozoa, a thorough evaluation of different extender solutions, cryoprotectants, and cooling and thawing rates are essential to develop optimum cryopreservation protocol for various species [3–5].

During the cryopreservation process, some factors may change the physiological status of sperm. The success of cryopreservation depends not only on preserving the motility of the spermatozoa but also on maintaining their metabolic functions [6]. Extender composition and cryoprotectant concentration are the main factors affecting cryopreservation success [7]. Extenders are required for dilution of fish sperm prior to cryopreservation and are generally designed to be compatible with the physiochemical composition of the fish seminal plasma. Most important function of the extenders is to maintain the spermatozoa in immotile state until required [8].

Cryoprotectants are added to the extenders to protect the cells against ice crystal formation during freezing and thawing [9]. Although cryoprotectants help to the prevention of cryoinjuries during freezing and thawing, they may become toxic to the cells when exposure time and concentration are increased [10, 11]. Thus, one of the most critical steps in successful cryopreservation of fish semen is the choice of the cryoprotectant and its ratio in the extender during the process.

Another important problem is the handling of sperm produced in small volumes by some fish species such as tilapia. In spite of packaging of sperm in traditional 0.25-mL and 0.5-mL straws has been successfully applied to freeze semen of the most fish species and to fertilize small egg batches [9], there is a lack of information regarding their usage in cryopreservation of Nile tilapia (*Oreochromis niloticus*) semen.

The Nile tilapia is one of the most cultivated freshwater fish species in the world aquaculture [12]. This species has great breeding potential due to its hardiness against worse environmental conditions, fast growth rate, adaptation to different environmental conditions (e.g. salinity, temperature), and also good organoleptic characteristics of its flesh [13, 14]. On the other hand, most of studies related with fish sperm cryopreservation have focused on some freshwater species, such as cyprinids [15, 16], salmonids [17, 18], catfishes [19, 20], and loach [21].

Even though many successes have been achieved in fish semen cryopreservation, the technique remains as a method that is difficult to standardize and use in all types of fishes. This is due to the fact that cryopreservation of sperm from different fish species required different conditions, where the protocol needs to be established individually [22]. To the best of our knowledge, there is limited information regarding cryopreservation of Nile tilapia sperm. In this concept, the effect of cryoprotectants and packaging methods on freezability and also on post-thaw quality of Nile tilapia sperm still remains unclear. Thus, standardization and simplification of cryopreservation procedure for Nile tilapia sperm are needed for commercial and gene bank applications.

The main aim of this study was to establish an efficient method for cryopreservation of Nile tilapia sperm that can be applied to aquaculture of this species. The present experiment was designed to study the straw volume (0.25 vs 0.5 mL) on Nile tilapia sperm quality after cryopreservation using glucose-Tris–based solution containing 10% dimethylacetamide.

2. Materials and methods

2.1. Reagents

The additives and other chemicals used in this study were obtained from local representative of Sigma-Aldrich Chemicals Company (St. Louis, MO, USA).

2.2. Broodstock handling

The experiments were carried out spawning season of the Nile tilapia. In the pre-spawning period, sexually mature male (n=15) and female (n=5) Nile tilapia were pit-tagged and kept separately in 150 L indoor tanks under constant environmental conditions. The broodstock tanks were provided with freshwater constantly at ratio of 1.5 L/s, while compressed air was provided trough air stones. The water temperature ranged from 27.2°C to 30.5°C, and salinity was maintained at 1.5 ppm. Nile tilapia was fed with floating pellets twice daily (1–5% body weight per day).

2.3. Gamete collection

Gametes were collected from healthy mature males and females following immersion anesthetization with 10 ppm quinaldine (Reanal Ltd., Budapest, Hungary) for a few minutes. For sperm collection, 1-mL tuberculin plastic syringe, without needle, was used to aspirate sperm released by gentle abdominal massage to eliminate urine in the ducts. Following, sperm samples were transferred individually into 1.5-mL Eppendorf tubes on ice (0–4°C). A 10-μL pipette tip connected to a mouth pipette was used to extract sperm cells, which were diluted 1:1 in Hanks' balanced salt solution (HBSS) (280 mOsm/kg, pH 7.0) in 1.5-mL microcentrifuge tubes and placed on ice until analysis. Eggs were also collected by gentle abdominal massaging and stored in HBSS at 25°C and used for fertilization within 30 min following stripping [23].

2.4. Gamete quality determination

One microlitre sperm of each sample was placed on a microscope slide and observed under a phase-contrast microscope (Olympus, Japan) at 100× magnification. The motility characteristics of the collected sperm samples were evaluated by adding activation solution (AS) (45 mM NaCl, 5 mM KCl, and 30 mM Tris–HCl, pH 8.2) [24] at a ratio of 1:100. Sperm cells that vibrated in place were considered as immotile. Only samples whose quality parameters ranging between the following values were used for the cryopreservation experiment: osmolarity 50–100 mOsm/kg, pH 7.0–8.0, and progressive motility 80–100% [25]. The quality of ova was determined from their morphological features seen under a dissecting microscope as described in the study of Fauvel et al. [26].

2.5. Cryopreservation and thawing experiments

Semen samples showing ≥80 motility was pooled into equal aliquots and chosen for cryopreservation experiments. Semen and extenders were kept at 4°C immediately under aerobic conditions prior to dilution. Pooled semen was diluted at a ratio of 1:3 with an extender composing 350 mM glucose and 30 mM Tris [27] containing 10% dimethylacetamide (DMA). The diluted semen were drawn into 0.25-mL or 0.5-mL plastic straws (IMV France) and sealed with polyvinyl alcohol (PVA). Following equilibration of semen for 10 min at 4°C, the straws were placed on a styrofoam rack floating on the surface of liquid nitrogen in a styrofoam box. Samples were frozen 3 cm above of the liquid nitrogen surface and exposed to the liquid nitrogen vapor (≈−140°C) for 10 min [28]. Following, frozen sperm cells were kept into the liquid nitrogen container (−196°C) until analyses for a few days.

The frozen sperm in different volume of straws were thawed in a water bath at 30°C for 20 s (0.25-mL straws) or at 30°C for 30 s (0.5-mL straws). Thawed semen was activated using activation solution (AS) (45 mM NaCl, 5 mM KCl, and 30 mM Tris–HCl, pH 8.2) [24] and observed under a phase-contrast microscope (Olympus, Japan) for progressive motility (%), progressive motility duration (s), and viability (%) evaluations (three replicates). At least five straws were used for each parameter evaluation with three replications.

2.6. Post-thaw sperm quality determination

The percent of motile spermatozoa and motility duration was immediately recorded following activation using a CCD video camera (CMEX-5, Netherland) mounted on a phase-contrast microscope (100×, Olympus BX43, Tokyo, Japan) at room temperature (20°C). Progressive spermatozoa motility and duration of progressive spermatozoa motility were evaluated from sperm with forward movement. Immotile spermatozoa were defined as spermatozoa that did not show forward movement after activation. Percentage of spermatozoa motility was determined within 30 s post-activation. Motility duration was evaluated by counting the time from spermatozoa activation until spermatozoa stopped moving. In order to assess viable sperm percentage, eosin-nigrosin preparations were made according to the method described by Bjorndahl et al. [29] and totally 300 sperm cells were counted on each slide at 1000× magnification. At least five straws were used for each evaluation parameter, and analyses were repeated three times for each treatment.

2.7. Fertilization experiments

Pooled eggs from five mature females were used to assess fertilization rates. In this stage, most of the HBSS was decanted from the eggs, and fertilization was carried out in dry Petri dishes (10 cm diameter). Fresh or thawed sperm was added over the eggs and gently mixed before activation with 20 mL of fertilization solution (3 g urea and 4 g NaCl in 1 L distilled water) [30]. Following fertilization process, 2 mL embryo buffer medium (EBM) (13.7 mM NaCl, 5.40 mM KCl, 0.25 mM Na_2HPO_4, 0.44 mM KH_2PO_4, 1.30 mM $CaCl_2$, 1.00 mM $MgSO_4$, 4.20 mM $NaHCO_3$ at 52 mOsm/kg and pH 7.0) was added for activation as described by Westerfield [31]. After 10 min, 100 mL EBM was added again over the eggs and was left undisturbed without movement in a convection type incubator at 27°C (Panasonic MCO-19M-PE, Japan). Unfertilized eggs were removed, and EBM was changed twice daily. After 48 h, the eggs were evaluated for fertilization results. Eggs that developed to stage 11 (embryonic keel and somite formation) were recorded as fertilized eggs, described by Galman and Avtalion [32].

Fertilization experiments were carried out using 1×10^5:1 spermatozoa/egg ratio with each straw types (0.25 or 0.5 mL) for the each aliquot of eggs (containing 100 eggs). Three straws were thawed for each fertilization treatment (three replications). For the control, another three aliquot of eggs (containing 100 eggs) were fertilized with fresh semen collected from two other males. Eggs were fertilized with fresh semen samples using the same number of sperm cells (1×10^5 cells) similar to treatments with frozen semen.

2.8. Statistical analysis

A two-way analysis of variance (ANOVA) including the straw volumes (0.25 and 0.5 mL) as fixed effects was used. Means were separated by Duncan's multiple range test and were considered at 5% level of significance. Results are presented as mean ± S.D. All analyses were carried out using SPSS 17 for Windows statistical software package.

3. Results

Semen volume was rather variable and ranged from 0.9 to 7.5 mL, with a mean volume of 3.6 ± 0.40 mL. Progressive motility was ranged from 60% to 90%, and mean motility was determined as 80.4 ± 0.15%. In addition, mean progressive spermatozoa motility duration (s),

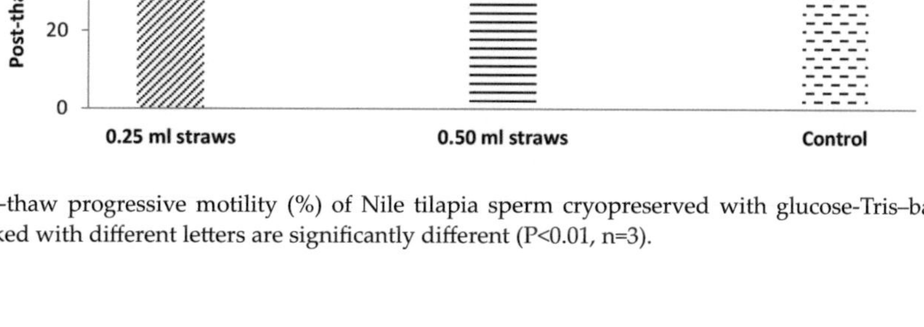

Figure 1. Post-thaw progressive motility (%) of Nile tilapia sperm cryopreserved with glucose-Tris–based extender. Columns marked with different letters are significantly different (P<0.01, n=3).

spermatozoa density (× 10^9/mL), viability (%), and pH values were determined as 64.2 ± 0.45 s, 1.75 × 10^9/mL, 92.5 ± 4.25%, and 7.2 ± 0.25, respectively. In addition, mean fertilization rate was determined with fresh semen, which was 72.5 ± 0.20%. The findings of the present study indicated that cryopreservation of sperm in glucose-Tris–based extender using 0.5-mL straws increased post-thaw progressive motility (**Figure 1**), duration of progressive motility (**Figure 2**), and fertility (**Figure 4**) (P<0.01). On the other hand, differences in terms of post-thaw cell viability were not significant among the treatments (**Figure 3**, P>0.01). The fertility of the frozen-thawed sperm showed high positive linear correlation with motility (r^2=1.000, **Figure 5**) and (r^2=0.9932, **Figure 6**) in case of using 0.25-mL and 0.5-mL straws.

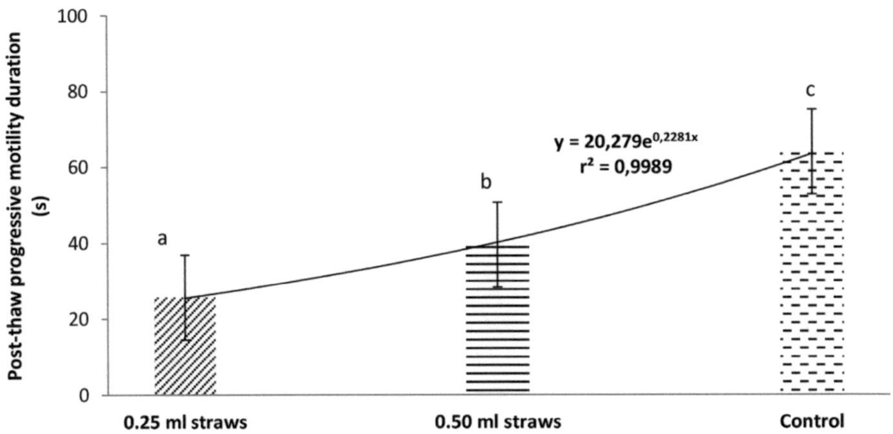

Figure 2. Post-thaw progressive motility duration (s) of Nile tilapia sperm cryopreserved with glucose-Tris–based extender. Columns marked with different letters are significantly different (P<0.01, n=3).

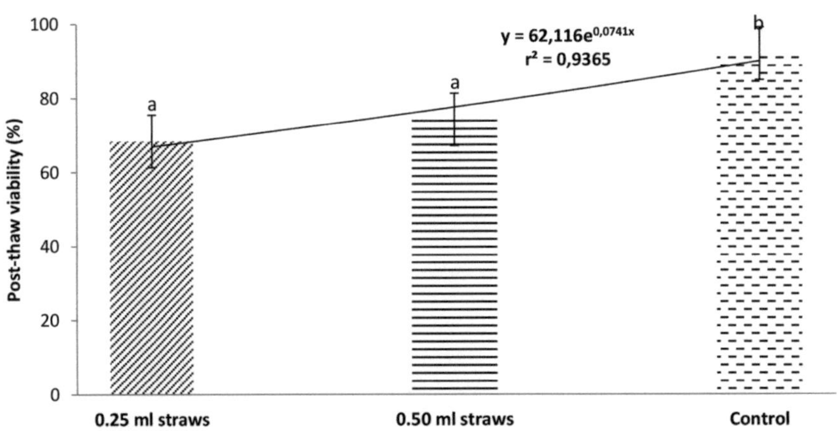

Figure 3. Post-thaw viability (%) of Nile tilapia sperm cryopreserved with glucose-Tris–based extender. Columns marked with different letters are significantly different (P<0.01, n=3).

http://dx.doi.org/10.5772/64835

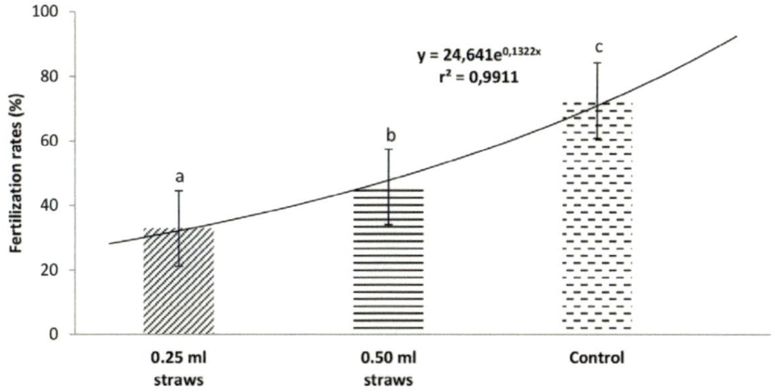

Figure 4. Post-thaw fertility (%) of Nile tilapia sperm cryopreserved with glucose-Tris–based extender. Columns marked with different letters are significantly different (P<0.01, n=3).

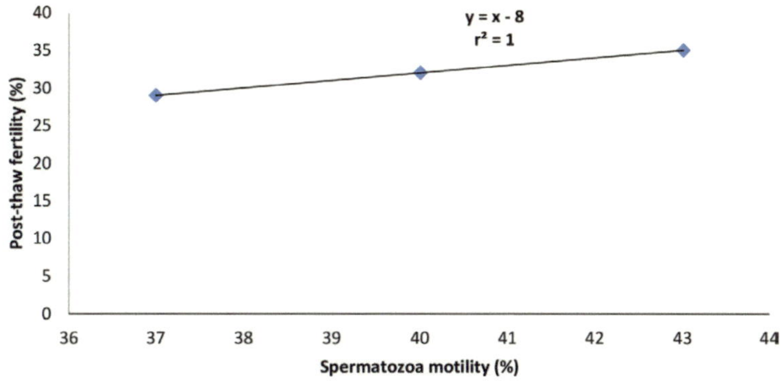

Figure 5. Relationship between post-thaw spermatozoa motility (%) and fertility (%) of Nile tilapia sperm cryopreserved with glucose-Tris–based extender using 0.25-mL straws.

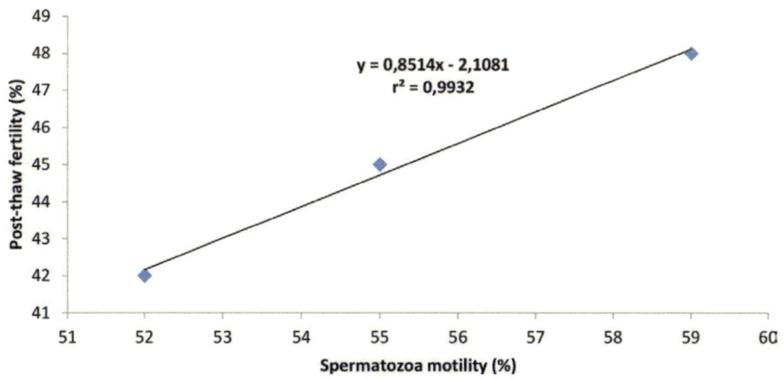

Figure 6. Relationship between post-thaw spermatozoa motility (%) and fertility (%) of Nile tilapia sperm cryopreserved with glucose-Tris–based extender using 0.50-mL straws.

4. Discussion

Tilapias are widely cultured in the tropical and subtropical regions of the world. Several species of tilapia are cultured commercially, but Nile tilapia is the predominant cultured species worldwide that its production reached 3,197,330 mt in 2012 [33].

Although Nile tilapia is a freshwater fish, it can tolerate a wide range of salinity [34]. Therefore, the expansion of its culture in sea and brackishwater has attracted the attention of fish farmers in recent years. However, limited reports have addressed semen cryopreservation in the Nile tilapia. Therefore, standardization and simplification of the cryopreservation procedure for the Nile tilapia sperm is needed for commercial application. On the other hand, because of limited amounts of data are available, comparison of the methods and results have been mainly made with the cyprinid species in this research. From this point of view, findings of this research significantly contribute improving of the protocols applied for the cryopreservation of the Nile tilapia sperm.

Sperm cryopreservation is an important biotechnological technique with specific advantages to the aquaculture industry. Improvements in semen cryopreservation techniques require in-depth knowledge of gamete physiology and the biochemical processes occurring during semen collection, processing, freezing, and thawing [35].

In spite of routinely using of cryopreserved semen in aquaculture artificial insemination programs worldwide, there are inconsistencies in experimental results [36]. The success of cryopreservation mainly depends on maintaining the spermatozoal metabolic functions [37]. The major factors affecting the results of insemination with frozen/thawed sperm are the type and properties of extenders and cryoprotectants, the damage caused by the formation of internal ice crystals due to the increase in solute concentration in the extension media and the interaction of these factors [38].

In the present study, glucose-Tris–based extender supplemented with DMA was used to cryopreserve Nile tilapia sperm. In spite of using glucose mainly as energetic substrate, it has been used due to its stabilization effects on the spermatozoa liposomal membrane [39]. It should be noted that, carbohydrate-based solutions such as glucose have also been found effective in some experiments [40, 41], and Tris is the most common buffer solution, not only for cyprinidae but also for other fish species [42].

Cryoprotectants are added to the extenders in order to protect the spermatozoa from damages during the freezing/thawing process. On the other hand, cryoprotectants can suppress most cryoinjuries when used at higher concentrations but at the same time it can become toxic to the cells. In addition, the amount and type of cryoprotectants used in sperm diluent depend on fish species and can affect the physiological and metabolic structure of the spermatozoa during cryopreservation procedure in different ways [43]. Therefore, selection of suitable type and concentration of the cryoprotectant is needed for the development of an effective cryopreservation protocol. However, comparison of different cryoprotectants and freezing/thawing protocols are difficult when each treatment tested for the ability of fertilization of the eggs by spermatozoa. For this reason, the protective effect of different cryoprotectants varies in different fish species. In this concept, several cryoprotectants have been mainly used for fish sperm cryopreservation, such as dimethyl sulphoxide (DMSO), dimethyl acetamide (DMA), glycerol (Gly), methanol (MeOH), ethylene glycol (EG), and propylene glycol (PG) [44–46].

Cryoprotectants are essential for preservation, but it is dissimilarly toxic to the cells. The toxicity tolerance level of the cells also depends on cryoprotectant concentration. Also, there are differences in permeability of the cells according to cryoprotectant types. In this study, DMA as penetrating cryoprotectant was used at 10% concentration, and diluted samples were equilibrated for 10 min at 4°C. Some authors recommend having an equilibration period following dilution, allowing cryoprotectants to penetrate the spermatozoa before cryopreservation [9, 47]. However, some authors reported equilibration process did not improve cryopreservation success in fish [48, 49].

On the other hand, the freezing conditions depended on the straw size and were also species specific. Insemination with cryopreserved semen of Arctic charr (*Salvelinus alpinus*) in 1.7-mL flat straws using 10% DMSO resulted highest percentage of eyed eggs (57.9 ± 11.6%) than with 0.5-mL and 2.5-mL straws [50]. No significant difference was obtained between fertilization percentage of blue catfish spermatozoa frozen in 0.5-mL and 1-mL straws. The larval hatch rates of striped trumpeter (*Latris lineata*) semen frozen in 0.25-mL and 0.5-mL straws (44.3 ± 2.9% and 44.2 ± 2.0%) [51]. In case of common carp, the use of conventional 0.5-mL straws resulted in 67 ± 17% hatching [24]. The findings of the present study demonstrated that progressive motility decreased for the two types of tested straws and varied from 55.7% in 0.5-mL to 40.2% in 0.25-mL straws. In addition, the 0.5-mL straws gave the best results in fertility as 45.7% when compared with 0.25-mL straws. The use of these small volumed straws during the artificial reproduction in fish can be time consuming, as many straws are needed to fertilize thousands of eggs. On the other hand, small volumed straws are more suitable for gene banking or fertilization of small amounts of eggs under laboratory conditions.

Successful fertilization of eggs using cryopreserved sperm is the final target of cryopreservation process. Fertilization ability of the cryopreserved sperm is a reliable approach to evaluate success of the cryopreservation protocol [52]. According to the results of the present study, Nile tilapia spermatozoa were influenced by cryopreservation process, and depending on this interaction, fertilization ability of frozen/thawed sperm decreased than fresh ones. The reason for the low fertility rate of frozen/thawed spermatozoa may be attributed to the changes in ultrastructural morphology, decrease in progressive motility and motility duration, and also possible toxic effects of the DMA.

Motility is one of the most important factors to assess fish sperm quality because it gives important information about the sperm cell's energy sources. In addition, better knowledge of the characteristics of fresh sperm motility is necessary to evaluate sperm quality in commercial hatcheries before artificial reproduction and also in laboratories before experiments. Spermatozoa motility is induced following releasing of the spermatozoa into the aquatic environment during natural reproduction or after transferring to an activation medium during controlled reproduction [53].

The observed decrease in sperm motility might be due to decrease in the percentage of sperm viability, high damage of sperm cells, or decrease in ATP content following cryopreservation. Similarly, Alavi et al. [54] determined that almost all studies on sturgeon sperm cryopreservation showed significant lower sperm motility and fertilizing ability of frozen/thawed sperm compared to that of the fresh sperm.

On the other hand, when fish spermatozoa are released into water or activation medium, they have a brief spermatozoal activity period [55]. For instance, in fresh sperm, the duration of spermatozoa motility in several cyprinids have been reported to last 120 s [56]. Similarly, in case of silver barb, the maximum motility period was observed until 150 s after water activation [4]. However, in case of frozen/thawed sperm, duration of mean post-thaw spermotozoa motility (32.0 ± 8.16 s) of the Nile tilapia was determined as lower than the results assessed with mirror carp [57] but higher than that of scaly carp [46] when DMSO, DMA, and glycerol were used as cryoprotectant. Similar results for the motility parameters of frozen-thawed spermatozoa were reported in fish in some experiments [48, 58, 59]. On the other hand, it is interesting to note that Godinho et al. [60] reported 241.2 ± 57.3 s post-thaw spermatozoa motility duration in glucose-based cryosolution containing 10% methanol in Nile tilapia.

In the present study, the applied sperm/egg ratio was 1×10^5:1 for fresh as well as frozen/thawed sperm, which probably resulted in excessive sperm concentrations in all batches. However, according to Lubzens et al. [48], the concentration of frozen/thawed sperm to be used to achieve optimal fertilization and hatching success is approximately 100 times higher than for fresh semen. This may be due to differences in extender compositions, cryoprotectant types, equilibration periods, egg quality, or applied protocols. In the present study, high positive correlation was determined between post-thaw spermatozoa motility and fertilization. This situation was consistent with the results that obtained from turbot (*Psetta maxima*) [61], common carp (*Cyprinus carpio*) [49], and African catfish (*Clarias gariepinus*) [62].

5. Conclusion

In conclusion, Nile tilapia sperm can be successfully cryopreserved in a glucose-Tris–based cryosolution containing 10% DMA with 0.5-mL straws. The applied protocol can be used in commercial hatcheries to facilitate artificial reproduction of Nile tilapia due to acceptable post-thaw motility and fertility results obtained. On the other hand, additional researches are needed to examine the growth and survival of the larvae originated from cryopreserved sperm.

Acknowledgements

This research was funded by the Mustafa Kemal University Scientific Research Fund (Project No: MKU BAP-384).

Author details

Yusuf Bozkurt[1*] and İlker Yavaş[2]

*Address all correspondence to: yfbozkurt@hotmail.com

1 Faculty of Marine Sciences and Technology, Department of Aquaculture, İskenderun Technical University, İskenderun, Hatay, Turkey

2 Faculty of Veterinary Medicine, Department of Reproduction and Artificial Insemination, Mustafa Kemal University, Antakya, Hatay, Turkey

References

[1] Daly J, Galloway D, Bravington W, Holland M, Ingram B: Cryopreservation of sperm from Murray cod, *Maccullochella peelii peelii*. Aquaculture. 2008; 285 (1-4):117-122. DOI: 10.1016/j.aquaculture.2008.08.023.

[2] Yildiz C, Bozkurt Y, Yavas I. An evaluation of soybean lecithin as an alternative to avian egg yolk in the cryopreservation of fish sperm. Cryobiology. 2013; 67 (1): 91-94. DOI: 10.1016/j.cryobiol.2013.05.008.

[3] Bozkurt Y, Yavas I, Karaca F. Cryopreservation of brown trout (*Salmo trutta macrostigma*) and ornamental koi carp (*Cyprinus carpio*) sperm. Current Frontiers in Cryopreservation, Edited by: Katkov, I. Section IV, p.293-304, Celltronix and Sanford-Burnham Institute for Medical Research USA, ISBN: 978-953-51-0302-8, 2012. 462p.

[4] Routray P, Dash SN, Dash C, Swain P, Sarkar SK, Sarangi N. Cryopreservation of silver barb *Puntius gonionotus* (Bleeker) spermatozoa: effect of extender composition, cryoprotective agents and freezing rate on their post-thawing fertilization ability. Aquaculture Research. 2008; 39 (15): 1597-1605. DOI: 10.1111/j.1365-2109.2008.02031.x.

[5] Bozkurt Y, Yavas I, Yildiz C. Effect of different avian egg yolk types on fertilization ability of cryopreserved common carp (*Cyprinus carpio*) spermatozoa. Aquaculture International. 2014; 22 (1): 131-139. DOI: 10.1007/s10499-013-9728-4.

[6] Watson PF. The causes of reduce fertility with cryopreserved semen. Animal Reproduction Science. 2000; 60-61: 481-492. DOI: 10.1016/S0378-4320(00)00099-3.

[7] De Graaf JD, Berlinsky DL. Cryogenic and refrigerated storage of rainbow smelt (*Osmerus mordax*) spermatozoa. Journal of the World Aquaculture Society. 2004; 35 (2): 209-216. DOI: 10.1111/j.1749-7345.2004.tb01076.x.

[8] Omitogun OG, Olaniyan OF, Oyeleye OO, Ojiokpota C, Aladele SE, Odofin WT. Potentials of short term and long term cryopreserved sperm of African giant catfish

(*Clarias gariepinus* Burchell 1822) for aquaculture. African Journal of Biotechnology. 2010; 9 (41): 6973-6982. DOI: 10.5897/AJB09.1680.

[9] Lahnsteiner F, Weismann T, Patzner RA. Cryopreservation of semen of the grayling (*Thymallus thymallus*) and the Danube salmon (*Hucho hucho*). Aquaculture. 1996a; 144 (1-3): 265-274. DOI: 10.1016/S0044-8486(96)01308-7.

[10] Bozkurt Y, Akcay E, Tekin N, Secer S. Effect of freezing techniques, extenders and cryoprotectants on the fertilization rate of frozen rainbow trout (*Oncorhynchus mykiss*) sperm. The Israeli Journal of Aquaculture-Bamidgeh. 2005; 57 (2): 125-130. DOI: 10524/19138.

[11] Christensen JM, Tiersch TR. Cryopreservation of channel catfish sperm: effect of cryoprotectant, straw size and extender formulation. Theriogenology. 2005; 63 (8): 2103-2112.DOI: 10.1016/j.theriogenology.2004.08.013.

[12] FAO. National aquaculture sector overview - Costa Rica. 2016. Available from: http://www.fao.org/fishery/countrysector/naso_costarica/en. [Accessed: 2016-05-20].

[13] Bhujel RC. A review of strategies for the management of Nile tilapia (*Oreochromis niloticus*) broodfish in seed production systems, especially hapa-based systems. Aquaculture. 2000; 181 (1-2): 37-59. DOI:10.1016/S0044-8486(99)00217-3.

[14] Little DC, Hulata G. Strategies for tilapia seed production. In: Beveridge, M.C.M., McAndrew, B.J. (Eds.), Tilapias: Biology and Exploitation. Fish and Fisheries Series, 25. Kluwer Academic Publishers, Dordrecht, The Netherlands, 2000. p. 267-326.

[15] Babiak I, Glogowski J, Brzuska E, Adamek J. Cryopreservation of sperm of common carp, *Cyprinus carpio* L. Aquaculture Research. 1997; 28 (7): 567-571. DOI: 10.1046/j. 1365-2109.1997.00897.x.

[16] Lahnsteiner F, Berger B, Horvath A, Urbanyi B, Weismann T. Cryopreservation of spermatozoa in cyprinid fishes. Theriogenology. 2000; 54 (9): 1477-1498. DOI:10.1016/ S0093-691X(00)00469-6.

[17] Conget P, Fernandez M, Herrera G, Minguell JJ. Cryopreservation of rainbow trout (*Oncorhynchus mykiss*) spermatozoa using programmable freezing. Aquaculture. 1996; 143 (3-4): 319-329. DOI:10.1016/0044-8486(96)01275-6.

[18] Cabrita E, Robles V, Alvarez R, Herráez MP. Cryopreservation of rainbow trout sperm in large volume straws: application to large scale fertilization. Aquaculture. 2001; 201 (3-4): 301-314. DOI:10.1016/S0044-8486(01)00636-6.

[19] Christensen JM, Tiersch TR. Cryopreservation of channel catfish spermatozoa: effect of cryoprotectant, straw size, and formulation of extender. Theriogenology. 1997; 47 (3): 639-645. DOI:10.1016/S0093-691X(97)00022-8.

[20] Viveiros ATM, So N, Komen J. Sperm cryopreservation of African catfish, *Clarias garepinus*: cryoprotectants, freezing rates and sperm:egg dilution ratio. Theriogenology. 2000; 54 (9): 1395-1408. DOI: 10.1016/S0093-691X(00)00462-3.

[21] Kopeika J, Kopeika E, Zhang T, Rawson DM, Holt WV. Detrimental effects of cryopreservation of loach (*Misgurnus fossilis*) sperm on subsequent embryo development are reversed by incubating fertilised eggs in caffeine. Cryobiology. 2003; 46 (1): 43-52. DOI: 10.1016/S0011-2240(02)00162-1.

[22] Chew PC, Zulkafli AR. Sperm cryopreservation of some freshwater fish species in Malaysia. Current Frontiers in Cryopreservation, Edited by: Katkov, I. Section IV, p. 269-292, Celltronix and Sanford-Burnham Institute for Medical Research USA, ISBN: 978-953-51-0302-8, 2012. 462p.

[23] Paleo GA, Godke RR, Tiersch TR. Intracytoplasmic sperm injection using cryopreserved, fixed, and freeze-dried sperm in eggs of Nile tilapia. Marine Biotechnology. 2005; 7 (2): 104-111. DOI:10.1007/s10126-004-0162-5.

[24] Horvath A, Miskolczi E, Urbányi B. Cryopreservation of common carp sperm. Aquatic Living Resources. 2003; 16 (5): 457-460. DOI:10.1016/S0990-7440(03)00084-6.

[25] Chao NH, Chao WC, Liu KC, Liao IC. The properties of tilapia sperm and its cryopreservation. Journal of Fish Biology. 1987; 30 (2): 107-118. DOI: 10.1111/j.1095-8649.1987.tb05737.x.

[26] Fauvel C, Omnes MH, Suquet M, Normant Y. Reliable assessment of overripening in turbot (*Scophthalmus maximus*) by a simple pH measurement. Aquaculture. 1992; 117 (1-2): 107-113. DOI:10.1016/0044-8486(93)90127-K.

[27] Horvath A, Miskolczi E, Mihalffy S, Osz K, Szabo K, Urbanyi B. Cryopreservation of common carp (*Cyprinus carpio*) sperm in 1.2 and 5 mL straws and occurrence of haploids among larvae produced with cryopreserved sperm. Cryobiology. 2007; 54 (3): 251-257. DOI:10.1016/j.cryobiol.2007.02.003.

[28] Tekin N, Secer S, Akcay E, Bozkurt Y. Cryopreservation of rainbow trout (*Oncorhynchus mykiss*) semen. The Israeli Journal of Aquaculture-Bamidgeh. 2003; 55 (3): 208-212. DOI: 10524/19084.

[29] Bjorndahl L, Soderlund I, Kvist U. Evaluation of the one-step eosin-nigrosin staining technique for human sperm vitality assessment. Human Reproduction. 2003; 18 (4): 813-816. DOI:10.1093/humrep/deg199.

[30] Yildiz C, Yavas I, Bozkurt Y, Aksoy M. Effect of cholesterol-loaded cyclodextrin on cryosurvival and fertility of cryopreserved carp (*Cyprinus carpio*) sperm. Cryobiology. 2015; 70 (2): 190-194. DOI:10.1016/j.cryobiol.2015.01.009.

[31] Westerfield M. The Zebrafish Book. Guide for the Laboratory Use of Zebrafish (*Danio rerio*). 1995. 3rd ed. University of Oregon Press, Eugene.

[32] Galman OR, Avtalion RR. Further study of the embryonic development of *Oreochromis niloticus* (*Cichlidae,* Teleostei) using scanning electron microscopy. Journal of Fish Biology. 1989; 34 (5): 653-664. DOI: 10.1111/j.1095-8649.1989.tb03347.x

[33] FAO. The state of world fisheries and aquaculture. 2014. Available from: http://www.fao.org/3/a-i3720e.pdf. [Accessed: 2016-05-20].

[34] Ernst DH, Watanabe WO, Ellingson LJ, Wicklund RI, Olla BL. Commercial-scale production of florida red tilapia seed in low- and brackish-salinity tanks. Journal of the World Aquaculture Society. 1991; 22 (1): 36-44. DOI: 10.1111/j.1749-7345.1991.tb00714.x.

[35] Bansal AK, Bilaspuri GS. Impacts of oxidative stress and antioxidants on semen functions. Veterinary Medicine International. 2011; (2011):686137. DOI: 10.4061/2011/686137.

[36] Holt WV. Alternative strategies for the long term preservation of spermatozoa. Reproduction Fertility and Development. 1997; 9 (3): 309-320. DOI:10.1071/R96082.

[37] Ciereszko A, Toth GP, Christ SA, Dabrowskila K. Effect of cryopreservation and theophylline on motility characteristics of lake sturgeon (*Acipenser fulvescens*) spermatozoa. Theriogenology. 1996; 45 (3): 665–672. DOI:10.1016/0093-691X(95)00412-2.

[38] Tekin N, Secer S, Akcay E, Bozkurt Y, Kayam S. Effects of glycerol additions on post-thaw fertility of frozen rainbow trout sperm, with an emphasis on interaction between extender and cryoprotectant. Journal of Applied Ichthyology. 2007; 23 (1): 60-63. DOI: 10.1111/j.1439-0426.2006.00792.x.

[39] Quinn PJ. A lipid-phase separation model of low-temperature damage to biological membranes. Cryobiology. 1985; 22 (2): 128-146. DOI:10.1016/0011-2240(85)90167-1.

[40] Billard R, Cosson MP. Measurement of sperm motility in trout and carp. In: Aquaculture (A Biotechnology in Progress), European Aquaculture Society, Breden, Belgium, 1989. p: 499-503.

[41] Babiak I, Glogowski J, Goryczko K, Dobosz S, Kuzminski H, Strzezek J, Demianowicz W. Effect of extender composition and equilibration time on fertilization ability and enzymatic activity of rainbow trout frozen spermatozoa. Theriogenology. 2001; 56 (1): 177-192. DOI:10.1016/S0093-691X(01)00553-2.

[42] Drokin S, Kopeika EF, Grishchenko VI. Differences in the resistance to cryopreservation and specifics of lipid compositions of spermatozoa of marine and freshwater fish species. Doklady Biochemistry 1989; 304 (1-6): 65-68.

[43] Stoss J, Holtz W. Cryopreservation of rainbow trout sperm. IV. The effect of DMSO concentration and equilibration time on sperm survival, sucrose and KCl as extender components and the osmolality of the thawing solution. Aquaculture. 1983; 32 (3-4): 321- 330. DOI:10.1016/0044-8486(83)90229-6.

[44] Routray P, Verma DK, Sarkar SK, Sarangi N. Recent advances in carp seed production and milt cryopreservation. Fish Physiology and Biochemistry. 2007; 33 (4): 413-427. DOI: 10.1007/s10695-007-9159-0.

[45] Bozkurt Y, Yavas I, Ogretmen F, Sivasligil B, Karaca F. Effect of glycerol on fertility of cryopreserved grass carp (*Ctenopharyngodon idella*) sperm. The Israeli Journal of Aquaculture-Bamidgeh. 2011; IIC:63: 635.

[46] Yavas I, Bozkurt Y, Yildiz C. Cryopreservation of scaly carp (*Cyprinus carpio*) sperm: effect of different cryoprotectant concentrations on post-thaw motility, fertilization and hatching success of embryos. Aquaculture International. 2014; 22 (1): 141-148. DOI: 10.1007/s10499-013-9698-6.

[47] Zhang X, Liu Y. Study of cryopreservation of fish spermatozoa. Acta Scientiarum Naturalium Universitatis Normalis Hunanensis 1991; 14: 255-259.

[48] Lubzens E, Daube N, Pekarsky I, Magnus Y, Cohen A, Yusefovich F, Feigin P. Carp (*Cyprinus carpio* L.) spermatozoa cryobanks–strategies in research and application. Aquaculture. 1997; 155 (1-4): 13-30. DOI:10.1016/S0044-8486(97)00106-3.

[49] Linhart O, Rodina M, Cosson J. Cryopreservation of sperm in common carp (*Cyprinus carpio*): sperm motility and hatching success of embryos. Cryobiology. 2000; 41 (3): 241-250. DOI:10.1006/cryo.2000.2284.

[50] Richardson GF, Miller TL, McNiven MA. Cryopreservation of Arctic charr, *Salvelinus alpinus* L., semen in various extenders and in three sizes of straw. Aquaculture Research. 2000; 31 (3): 307-315.

[51] Ritar AJ. Artificial insemination with cryopreserved semen from striped trumpeter. Aquaculture. 1999; 180: 177-187.

[52] Lemma A. Effect of cryopreservation on sperm quality and fertility. In: Artificial Insemination in Farm Animals. Milad Manafi (Ed.), 2011. p. 191-216. ISBN: 978-953-307-312-5.

[53] Alavi SMH, Cosson J. Sperm motility in fishes. II. Effects of ions and osmolality: A review. Cell Biology International. 2006; 30 (1): 1-14. DOI: 10.1016/j.cellbi.2005.06.004.

[54] Alavi SMH, Hatef A, Psenicka M, Kaspar V, Boryshpolets S, Dzyuba B, Cosson J, Bondarenko V, Rodina M, Gela D, Linhart O. Sperm biology and control of reproduction in sturgeon: (II) sperm morphology, acrosome reaction, motility and cryopreservation. Reviews in Fish Biol Fisheries. 2012; 22 (4): 861-886. DOI: 10.1007/s11160-012-9270-x.

[55] Morisawa S, Morisawa M. Acquisition of potential for sperm motility in rainbow trout and chum salmon. Journal of Experimental Biology. 1986; 126: 89-96.

[56] Suzuki R. Sperm activation and aggregation during fertilization in some fishes. III. Non-species specificity of stimulating factor. Annotations of Zoology of Japan. 1959; 32: 105-111.

[57] Akcay E, Bozkurt Y, Secer S, Tekin N. Cryopreservation of mirror carp semen. Turkish Journal of Veterinary and Animal Sciences. 2004; 28 (5): 837-843.

[58] Kurokura H, Hirano R, Tomita M, Iwahashi M. Cryopreservation of carp sperm. Aquaculture. 1984; 37 (3): 267-273. DOI: 10.1016/0044-8486(84)90159-5.

[59] Lahnsteiner F, Berger B, Weismann T, Patzner RA. Motility of spermatozoa of *Alburnus alburnus* (Cyprinidae) and its relationship to seminal plasma composition and sperm metabolism. Fish Physiology and Biochemistry. 1996b; 15 (2): 167-179. DOI: 10.1007/BF01875596.

[60] Godinho HP, Amorim VMC, Peixoto MTD. Cryopreservation of the semen of Nile tilapia *Oreochromis niloticus*, var. Chitralada: cryoprotectants, spermatozoa activating solutions and cryogenic refrigerator. Revista Brasileira de Zootecnia. (2003); 32: 6(1), 1537-1543. DOI: 10.1590/S1516-35982003000700001.

[61] Dreanno C, Cosson J, Suquet M, Seguin F, Dorange G, Billard R. Nucleotides content, oxidative phosphorylation, morphology and fertilizing capacity of turbot (*Psetta maxima*) spermatozoa during the motility period. Molecular Reproduction and Development. 1999; 53 (2): 230-243. DOI: 10.1002/(SICI)1098-2795(199906)53:2<230::AID-MRD12>3.0.CO;2-H.

[62] Rurangwa E, Volckaert FAM, Huyskens G, Kime DE, Ollevier F. Quality control of refrigerated and cryopreserved semen using computer-assisted sperm analysis (CASA), viable staining and standardized fertilisation in African catfish (*Clarias gariepinus*). Theriogenology. 2001; 55 (3): 751-769. DOI:10.1016/S0093-691X(01)00441-1.

Chapter 6

Long-Term Storing of Frozen Semen at −196°C does not Affect the Post-Thaw Sperm Quality of Bull Semen

Alfredo Ramírez-Reveco, Jorge Luis Hernández and Pablo Aros

Additional information is available at the end of the chapter

http://dx.doi.org/10.5772/64948

Abstract

Today, it is theoretically assumed that frozen storage of semen doses in liquid nitrogen guarantees sperm functionality indefinitely. However, there are few studies that objectively evaluate the effects of long-term storage on sperm quality parameters. In this study, we show a freezability analysis of bull semen stored for 1, 10, 25, 40 and 45 years at −196°C. Sperm viability and full sperm motility were analyzed by CASA system, and acrosome integrity was assessed with Coomassie blue staining. Our results showed that sperm viability and total sperm motility were not affected by long-term cryopreservation at −196°C. Specifically, we did not find any significant differences ($p > 0.05$) associated between different long-time storing analyzed; both parameters showed optimal values of sperm viability and total sperm motility (both over 60%). Additionally, the acrosomal integrity parameter was not affected, showing an optimal range (87±1.6 - 95±0.5%). We conclude that the sperm quality of bovine semen is not affected by long-term storage at −196°C. However, future field trials will be necessary in order to validate that both fertility and embryo viability are maintained for the times analyzed.

Keywords: sperm, cryopreservation, storage, liquid nitrogen, freezability, sperm quality

1. Introduction

Cryopreservation protocols for the bull used in the animal production industry began in the 1950s [1]. Since then, both the packaging type and cryopreservation system were changing on the time. Primary considerations in the selection of a system for packaging semen were fertility, insemination preference, ease of handling, ease of identification, freedom from contamination, economics of storage and efficiency of ejaculates [2]. Sperm are commonly packaged in one of three ways: (a) glass ampoules, normally containing 0.5–1.2 ml of frozen semen; (b) pellets containing about 0.1 ml; and (c) polyvinyl chloride straws with a volume of 0.25–0.5 ml.

Early field trials showed that the bovine semen frozen to −79°C and packed on dry ice could still yield high fertility [3]. Regarding to time storage factor, studies of sperm motility have indicated a descent in sperm viability after storage [4, 5]. On the other hand, field trials carried out at the Reading Cattle Breeding Centre (Great Britain, 1960) indicated no effect on conception rate when using long-term semen stored in a dry ice alcohol mixture for 4 years [6].

Until the 1970s, it was thought that frozen semen could be indistinctly stored in mechanical freezers at about −25°C, in solid carbon dioxide at −79°C, or in liquid nitrogen at −196°C. However, an inverse relationship between preservation of sperm viability and storage temperature was shown [7]. Briefly, most frozen semen was stored in a mixture of dry ice and alcohol at −79°C, which ultimately decreased fertility [8–10]. Meanwhile, studies of frozen semen stored at −196°C have shown a consistently high non-return rate [11–14].

Furthermore, since the 1970s, there have been mentions that deterioration continues even when sperm are stored in liquid nitrogen; suspecting that aging of spermatozoa may occur if semen is stored for long periods of time, and this may be associated with embryo mortality and delayed return [9, 15].

Studies by Salisbury and Hart [16] suggested that bovine frozen sperm have a low fertility level and promote increased embryonic mortality after 18 months of storage at −196°C, but other studies have been unable to confirm this. In this context, Strom [14] found no evidence of reduced fertility when approximately 60,000 inseminations were performed with semen packaged in pellets, following storage in liquid nitrogen at −196°C for approximately 1–1.5 years. Cassou [17] reported no difference in fertility after 285,551 inseminations with frozen semen in straws were stored at −196°C for up to 4.5 years. Similarly, a field trials of Roettger et al [18], with 100,000 inseminations and using frozen semen stored at −196°C for 5 years, a normal fertility rate was evidenced.

Field trials using frozen semen packaged in ampoules, pellets and straws have indicated that the influence of packaging methods in fertility has been inconsistent [2]. Therefore, according to these authors, if the semen fertility stored in LN is reduced with time, regardless of packaging technique, some factors other than storage are responsible.

Today, cryopreservation in liquid nitrogen (−196°C) is a technique that allows for long-term storage of spermatozoa [19]. This is a highly practical method in breeding programs for domestic animals and is used to maintain the establishment and genetic diversity of gene banks [20, 21].

Cryopreservation requires many stages during cooling/freezing and thawing procedures, which interactively affect its success [22, 23]. On the other hand, it is assumed that storage period in deep freezing does not affect sperm viability [24, 25], and there is argument that

http://dx.doi.org/10.5772/64948

spermatozoa retain their fertilizing potential indefinitely when stored at −79°C in dry ice, or at −196°C in liquid nitrogen [26]. However, there is a scarcity of studies designed in order to detect a decrease in reproductive performance of cryopreserved semen as a function of storage time.

In this respect, Mazur [27] proposed that, by accumulative cosmic radiation, more than 3,000 years. Then the question arises, how long cells can be stored in liquid nitrogen without suffering damage? This question is probably irrelevant if the cellular storage temperature is below −120°C, where chemical reactions do not occur in a human timescale. Moreover, at −196°C, the thermally driven reactions only can occur on a geological timescale [28].

There is the possibility of slow accumulation of direct damage from ionizing radiation, but this becomes significant only after centuries of storage [29]. Yet, as previously mentioned, insemination trials with frozen stored semen [16] suggest a far shorter time period of optimal semen storage at the above-mentioned low temperatures, in consideration of fertility rate maintenance. However, other evidence suggests that this could be due to inadequate maintenance of temperatures [26]. There are two studies that strongly reinforce the idea that fertilization potential is maintained in long-term storage in liquid nitrogen. Specifically, the *in vitro* fertilization (FIV) was obtained using frozen spermatozoa, stored in nitrogen liquid during 37 and 27 years, for bovine and human, respectively [30, 31].

In relation to sperm quality parameters, there are few studies that objectively evaluate the effects of long-term sperm storage. Leibo et al. [30] reported a normal bovine sperm motility after 37 years, and Rofeim and Gilbert [32] reported no statistical reduction in human sperm quality after 5 years of follow-up. More recently, although it was not analyzed by computer-assisted sperm analysis (CASA system), Malik et al. [33] showed that the viability and motility of thawed sperm stored in liquid nitrogen during 6 years were lower than 1- to 2-year storage .

The main goal of this study was to assess, through CASA system, the main sperm quality parameters of cryopreserved and stored bull semen in liquid nitrogen for 10, 25, 40 and 45 years.

2. Materials and methodology

2.1. Seminal doses, race of donor and processing

In this study, a total of 75 commercial doses from bulls Friesian breed were used. The cryopreserved germplasm were defined and divided into five groups according to the storage time. For each group, 15 seminal doses from five different donors (three each) were considered. All seminal doses used were collected, processed, packaged, cryopreserved and stored (−196°C) using commercial standard procedures by Center of Artificial Insemination (CIA), belonAustral University Chile (UACh).

The mean storage times or groups of the semen doses analyzed were the following: 45, 40, 25 and 10 years, which were packaged in glass ampoules, pellets, short straws and fine straws, respectively. As a control, commercial frozen doses cryopreserved in fine straws and stored in liquid nitrogen for 1 year were used. The following thawing protocols were used, in accordance with cryopreservation packaging supports (that is showed in **Figure 1**): Ampoule samples were thawed in thermo-stated water bath at 50°C for 75 s; pellet samples were thawed in a Thermos-stated water bath at 40°C for 55 s; short straw (Mini-Tubes) and fine straw samples were thawed in a thermo-stated water bath at 37°C for 30 s.

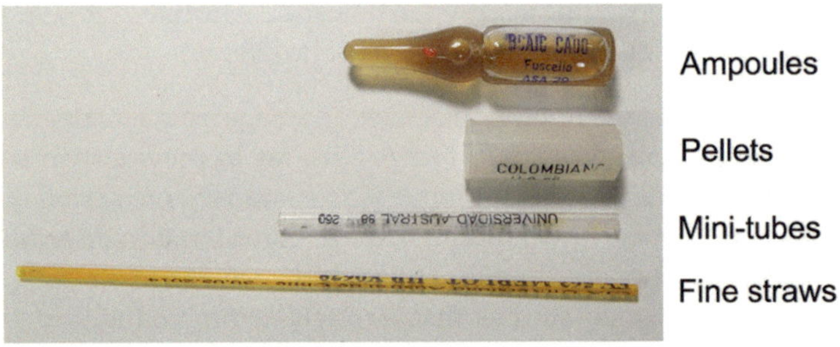

Figure 1. Different freezing packaging system used in this study.

2.2. Sperm quality analysis of seminal doses

2.2.1. Plasma membrane integrity (viability assessment)

Plasma membrane integrity was determined using the acridine orange (AO)/propidium iodide (PI) double-staining technique, according to Córdova et al. [34], with modifications. Briefly, post-thawing samples (3 μL) were mixed (1:1) with a staining aqueous solution composed of 20 μM AO and 10 μM PI in a tempered microscope slide. Stained samples were analyzed using the CASA System (viability module of the Sperm Class Analyser®, Microptic, Spain) coupled to an epifluorescence microscope (Nikon E200, Japan) with a high-velocity camera (Basler AG, Germany). Viability percentages were established from a minimum of 1000 spermatozoa for each sample.

2.2.2. Sperm motility assessment

Sperm motility was assessed using the CASA system (Motility module of the Sperm Class Analyser®, Microptic, Spain), according to Ramírez et al. [35], with modifications. A total of 6 μL aliquots of samples was then placed on a prewarmed (37°C) slide and covered with a 24 mm^2 coverslip. The motility analysis by CASA system was based on the analysis of 25 consecutive, digitalized photographic images taken over a time lapse of 1 s, obtained from a single field using a negative objective (10× magnification) and a phase contrast microscope (Nikon E200, Japan), coupled to a high-velocity camera (Basler AG, Germany, scA780 54tc). Four or seven separate fields were taken for each sample (at less 500 spermatozoa analyzed). Sperm motility parameters were as follows: curvilinear velocity (VCL); linear velocity (VSL); mean velocity (VAP); linearity coefficient (LIN): (VSL/VCL) × 100(%). Straightness coefficient

(STR): (VSL/VAP) × 100(%). Wobble coefficient (WOB): (VAP/VCL) × 100(%). Mean amplitude of lateral head displacement (ALH); frequency of head displacement (BCF). Bovine configuration of CASA system used was as follows—capture: 25 frames/s; particle area range: 5–70 μm^2; classification according to velocity (VAP): static < 10 μm/s < slow < 25 μm/s < medium < 50 μm/s < rapid. The progressive motility was defined as the percentage of spermatozoa showing an STR above 70%.

2.2.3. Sperm acrosomal integrity assessment

The structural status of sperm acrosomes was assessed using Coomassie G-250 staining, according to Larson et al. [36]. Briefly, sperm aliquots were washed in TBS, fixed and permeabilized for at least 30 min at 4°C in 100% methanol. Permeabilized spermatozoa dried onto slides were then covered with a droplet of staining solution (0.22% W/V Coomassie blue G-250; 50% methanol and 10% glacial acetic). The samples were washed with excess of bidistilled water, dried and observed under 100 × oil immersion lens. Percentage of stained cells was determined by counting of at less 300 spermatozoa.

2.3. Statistical analysis

Statistical analyses were performed using one-way ANOVA with post hoc Bonferroni multiple comparison tests. For the analysis, we used GraphPAD (Prism 6) software and differences were considered significant and highly significant for p values of <0.05 and <0.01, respectively.

3. Results and discussion

Our results showed that integrity of plasma membrane is not altered by long-term storage at −196°CC (**Figures 2A** and **3A**). Specifically, we did not find any significant differences associated with the different storage times analyzed (>10 years) or in relation to control ($p < 0.05$), with a post-thaw sperm viability percentage that oscillated between 60 ± 1.8 and 68 ± 2.1. A similar situation was observed in the acrosomal integrity analysis, wherein the only significant difference observed associated with storage time was specifically between 45 and 40 years ($p < 0.01$) (**Figures 2B** and **3C**). Despite these differences, the percentage of acrosomal integrity after thawing ranged between 87 ± 1.6 and 95 ± 0.5, and ultimately acceptable enough even for fresh semen. On the other hand, comparative analysis of total sperm motility did not show significant differences between times storage analyzed. Specifically, all values shown are above 60%, ranging between 60 ± 2.4 and 66 ± 3.2, similar to viability results (**Figure 2A** and **2C**).

In respect to progressive motility, a higher significant value was found in samples from frozen semen stored by 40 years ($p < 0.01$) (**Figures 2D** and **3B**). Additionally, a similar behaviour in other cinematic parameters (VSL, VAP, LIN and WOB) was observed in these samples, showing higher values ($p < 0.001$) (**Table 1**). Consistently, lowest values of hyperactivity, ALH and BCF were observed in the frozen semen stored for 40 years ($p < 0.001$). This great differences in progressive motility and other cinematic parameters in samples of seminal doses storage by

40 years may have been influenced by the thawing solution used in the pellets tube, particularly with the presence of sodium bicarbonate (30.9 mM) [37].

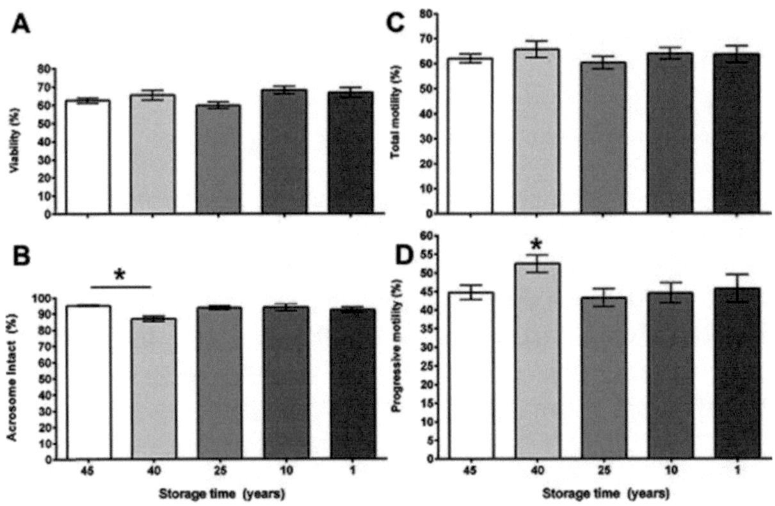

Figure 2. Freezability evaluation of frozen-thawed bovine sperm. A) Plasma membrane integrity or viability (%) analysis by CASA system using double stain (propidium iodide and acridine orange). A minimum of 1000 spermatozoa were counted for each assay. B) Acrosomal integrity (%) analysis by Coomassie G-250 staining. A minimum of 200 spermatozoa were counted for each assay. C) Total sperm motility (%) analysis by CASA System, VCL > 10 μm/s. A minimum of 500 spermatozoa were counted for each assay. D) Progressive sperm motility (%) analysis by CASA System, STR > 70%. A minimum of 500 spermatozoa were counted for each assay. Each bar (storage time) represents the mean ± SEM of a total of 15 doses were analyzed (from five bulls, three doses for each time analyzed). Significant differences among time storing are shown: one-way ANOVA/Bonferroni post-test ($p < 0.001$ or by different letters).

The results shown differ from those obtained by Malik et al. [33], who reported a significant decrease in both the viability and motility associated with prolonged storage (6 years versus 1–2 years). It is highly probable that these discrepancies, in the case of both viability and motility, are due to differences in the sensitivity of the technique used, nigrosin/eosin staining and bright field microscopy versus acridine orange/propidium iodine and epifluorescent microscopy in our case. This could also be due to the evident differences associated with the use of analysis of subjective sperm motility analysis versus our use of CASA system.

There is an argument that spermatozoa store at −79°C (in dry ice) or at −196°C (in liquid nitrogen) retain their fertilizing potential indefinitely [26]; however, the storage time results that studies are controversial. Effectively, although Mazur [29] proposed that several centuries are required, of liquid nitrogen storage, for that ionizing cosmic radiation alters or damages the DNA of the cell. However, there are studies that raise a discrepancy respect that cryopreservation, for several years, completely stops the processes of sperm biochemistry, but whose storage times do not exceed 6 years.

Fourteen five years ago, Salisbury and Hart [16] proposed that bovine frozen sperm have a low fertility level and promote increased embryonic mortality after 1.5 years of storage at −196°C. More recently, Haugan et al. [19] based on results of field trials indicated that the likelihood of conception decreased only a little more than one percentage after 5.5 years of storage, but that level of decline seems to be so important because the calving rate predicted by multiple logistic regression was 59.2%, optimal value according to commercial standard for frozen semen. Contrary, field trial results of Strom [14] found no evidence of reduced fertility when was used frozen semen storage by 1–1.5 years. Additionally, Cassou [17] and Roettger et al [18] reported no difference in fertility when were used frozen semen stored at −196°C for up to 4.5 and 5 years, respectively. Unfortunately, there are no field trials in that and both pregnancy and calving rates have been analyzed; to rule out or confirm effects of prolonged storage on embryo mortality, this would be the only one way to resolve the question.

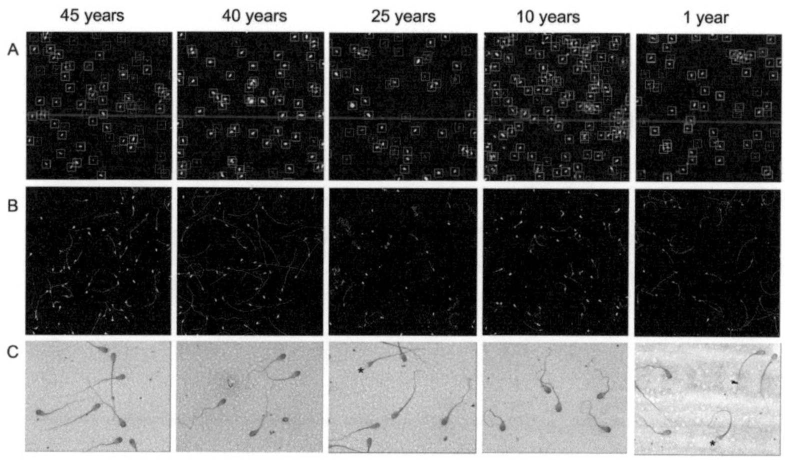

Figure 3. Representative's field captures for sperm quality parameters analyzed. (A) Plasma membrane integrity analysis, green or red fluorescents marks correspond to sperm recognized as live or dead, respectively. Images obtained with epifluorescence microscope, objective: 10×. (B) Sperm motility analysis, the tracking in red, green, blue and yellow, correspond to sperm sorted according velocity: rapid, medium, slow and static, respectively. Images obtained with phase contrast microscope, objective: 10×-negative. (C) Acrosomal integrity analysis, Coomassie blue G-250 stained acrosome-intact sperm or those with acrosome reacted or damaged (asterisks). Images obtained with bright field microscope, objective: 40×.

Storage time at −196°C	VCL (μ/s)	VSL (μ/s)	VAP (μ/s)	LIN (%)	SRT (%)	WOB (%)	ALH (μ)	BCF (Hz)
45 years	68.23 ± 1.8^{a}	34.88 ± 1.5^{a}	43.94 ± 1.6^{a}	47.8 ± 1.3^{a}	71.5 ± 1.3^{a}	63.1 ± 0.9^{a}	2.57 ± 0.0^{a}	9.38 ± 0.29^{a}
40 years	77.27 ± 2.5^{a}	52.63 ± 2.1^{b}	65.47 ± 2.9^{b}	60.2 ± 1.6^{b}	72.5 ± 1.4^{a}	78.4 ± 1.8^{b}	1.97 ± 0.1^{b}	7.07 ± 0.6^{b}
25 years	75.67 ± 2.1^{a}	32.46 ± 1.4^{a}	48.26 ± 1.9^{a}	42.5 ± 1.3^{a}	63.9 ± 1.3^{b}	63.0 ± 1.2^{a}	2.81 ± 0.0^{a}	9.46 ± 0.3^{a}
10 years	78.64 ± 2.8^{a}	34.03 ± 2.1^{a}	50.23 ± 2.2^{a}	41.7 ± 2.1^{a}	63.0 ± 2.0^{b}	62.2 ± 1.6^{a}	2.94 ± 0.1^{a}	8.87 ± 0.2^{a}
1 year	69.08 ± 5.0^{a}	32.42 ± 2.3^{a}	45.02 ± 3.2^{a}	45.0 ± 2.1^{a}	$66.9 \pm 2.2^{a,b}$	63.5 ± 1.3^{a}	2.59 ± 0.1^{a}	9.19 ± 0.4^{a}

Values are expressed as mean ± SEM. Different superscript letters (a and b) indicate significant differences among storage times (one-way ANOVA, $p < 0.001$).

Table 1. Effect of storage time at −196°C on bull sperm kinetic parameters.

Our result showed that the more important parameters of sperm quality non-present changes associated to storage times analyzed (1–10–25–45 years). Considering that the plasma and acrosomal membrane integrity are two irreversible parameters of sperm quality, and that the motility is commonly believed to be one of the most important characteristics associated with the fertilizing ability of semen [38]. Our freezability data, analyzed as a whole, suggest that fertilizing potential of the seminal dose is commercially analyzed, independent of storage time, and it is high. In this respect, Budworth et al. [39, 40] observed significant correlation of the sperm motility and sperm velocity with the competitive fertility index. Moreover, Amann [41] reported a high level of correlation between competitive fertility index and sperm motility, VCL, VSL parameters, with 0.80, 0.68 and 0.70, respectively.

We conclude, and categorically, that the basic parameters of sperm quality of bovine semen are not affected by long-term storage at −196°C. Complementary analysis, including other aspects as to mitochondrial metabolism, reactive oxygen species (ROS) levels, DNA fragmentation and chromatin integrity, could shed light on possible and potential changes induced for prolonged storage.

Future studies of embryo production by *in vitro* fecundation (IVF) and field trials are needed, in order to confirm effects associated to long-term sperm storing at −196°C on fertility, embryonic viability and calving rate.

Acknowledgements

Thanks to Research and Development Direction from Austral University of Chile, by support to Laboratory of Cryobiology and Spermatic Function Analysis. Special thanks to all professional and technical personnel of the Center of Artificial Insemination, that from the 1960s set up and development the bovine semen cryopreservation in Chile and whose seminal doses stored for long term has made this study possible.

Author details

Alfredo Ramírez-Reveco[1*], Jorge Luis Hernández[2] and Pablo Aros[1,3]

*Address all correspondence to: alfredoramirez@uach.cl

1 Laboratory of Cryobiology and Spermatic Function Analysis, Animal Science Institute, Faculty of Veterinary Science, Austral University of Chile, Valdivia, Chile

2 Center of Artificial Insemination, Faculty of Veterinary Science, Austral University of Chile, Valdivia, Chile

3 School of Veterinary Medicine, Faculty of Veterinary Science, Austral University of Chile, Valdivia, Chile

http://dx.doi.org/10.5772/64948

References

[1] Walters EM, Benson JD, Woods EJ & Critser JK. The history of sperm cryopreservation. In: Pacey, A.A., & Tomlinson, M.J., editors. Sperm banking: theory and practice. Cambridge: Cambridge University Press; 2009; p. 1–17.

[2] Berndtson WE & Pickett BW. Techniques for the cryopreservation and field handling of bovine spermatozoa. In: The integrity of frozen spermatozoa. Proceeding of a Round-table Conference held on 6–7 April 1976. National Academy of Sciences Washington, DC. 1976; p. 53–77.

[3] Bratton RW, Foote RH & Cruthers JC. Preliminary fertility results with frozen bovine spermatozoa. Journal of Dairy Science. 1955; 38 (1): 40–46.

[4] Dunn HO & Hafs HD. Extenders and techniques for freezing bovine spermatozoa. Journal of Dairy Science. 1953; 36 (6): 577–577.

[5] Buch NC, Smith VR & Tyler WJ. Bull and line differences in the survival of spermatozoa after freezing and thawing. Journal of Dairy Science. 1956; 39: 1712–1716.

[6] Melrose DR. Ultra-low temperature storage of sperm. In: Maule, JP., editors. The semen of animals and artificial insemination. 1st ed. Bucks, England: Commonwealth Agricultural Bureaux; 1962; p. 109–132.

[7] Gomes WR. Artificial insemination. In: Cole, HH., & Cupps, PT., editors. Reproduction in domestic animals. 3rd ed. London: Academic Press Inc; 1977; p. 257–284.

[8] Mixner JP & Wiggin SH. Stored frozen semen fertility. AI Digest. 1960; 8 (7): 19.

[9] Salisbury GW & Flerchinger FH. Aging phenomena in spermatozoa. I. Fertility and prenatal losses with use of liquid semen. Journal of Dairy Science. 1967; 50 (10): 1675–1678.

[10] Stewart DL. Observation on the fertility of frozen semen stored at −79°C and −196°C. In: In Proceedings of International Congress on Animal Reproduction and Artificial Insemination; 1964; Vol. 4. p. 617.

[11] Clegg EO & Pickett BW. Effect of storage at −196°C on fertility. AI Digest. 1966; 14 (3): 12–13.

[12] Foote RH. Aging of spermatozoa during storage in liquid nitrogen. In: In Proceedings of the Fourth NAAB Technical Conference on Artificial Insemination and Reproduction. 1972; Vol. 28.

[13] Lindström U. Pellets: experiences and results of storage over 3 years. In: VII International Congress on Animal Reproduction and Artificial Insemination, Munich. 1972; p. 1428–1435.

[14] Strom B. Influence upon fertility of bull semen of storage time in liquid nitrogen. In: International Congress on Animal Reproduction and Artificial Insemination [Proceedings]. 1968; p. 1171–1174.

[15] Salisbury GW. Effect of season and storage at –79°C to -88°C on fertility and prenatal losses. Journal of Dairy Science. 1967; 50 (10): 1683–1689.

[16] Salisbury GW & Hart RG. Gamete aging and its consequences. Biology of Reproduction. 1970; (2): 1–13.

[17] Cassou R. A technique for semen preservation and its application to breed improvement programmes. AI Co-op Centre. 1972; 61-L'Aigle, France. Personal publication (Cited by Berndtson WE & Pickett BW, 1976) [2].

[18] Roettger LW, Salisbury GW, Lee AJ, Boyd LJ & Ingalls W. *In vitro* aging of frozen bull semen. Journal of Dairy Science. 1975; 58 (5): 767–768.

[19] Haugan T, Gröhn YT, Kommisrud E, Ropstad E & Reksen O. Effects of sperm concentration at semen collection and storage period of frozen semen on dairy cow conception. Animal Reproduction Science. 2007; 97 (1): 1–11.

[20] Saint JM, Lecoq R, Seigneurin F, Blesbois E & Plouzeau E. Cryopreservation of semen from endangered pheasants: the first step towards a cryobank for endangered avian species. Theriogenology. 2003; 59 (3): 875–888.

[21] Shivaji S, Kholkute SD, Verma SK, Gaur A, Umapathy G, Singh A, Sontakke S, Shailaja H, Reddy A, Monika S, Sivaram V, Jyotsna B, Bala S, Ahmed SM, Bala A, Chandrashekar BVN, Gupta S, Prakash S & Singh L. Conservation of wild animals by assisted reproduction and molecular marker technology. Indian Journal of Experimental Biology. 2003; 41: 710–723.

[22] Medeiros CMO, Forell F, Oliveira ATD & Rodrigues JL. Current status of sperm cryopreservation: why isn't it better? Theriogenology. 2002; 57 (1): 327–344.

[23] Neild DM, Gadella BM, Chaves MG, Miragaya MH, Colenbrander B & Agüero A. Membrane changes during different stages of a freeze–thaw protocol for equine semen cryopreservation. Theriogenology. 2003; 59 (8): 1693–1705.

[24] Hammerstedt RH, Graham JK & Nolan JP. Cryopreservation of mammalian sperm: what we ask them to survive. Journal of Andrology. 1990; 11 (1): 73–88.

[25] Watson PF. Recent developments and concepts in the cryopreservation of spermatozoa and the assessment of their post-thawing function. Reproduction, Fertility and Development. 1995; 7 (4): 871–891.

[26] Vishwanath R & Shannon P. Storage of bovine semen in liquid and frozen state. Animal Reproduction Science. 2000; 62 (1): 23–53.

http://dx.doi.org/10.5772/64948

[27] Mazur P. Fundamental aspects of the freezing of cells with emphasis on mammalian ova and embryos. In: Proceedings of 9th International Congress on Animal Reproduction. A.I. Madrid, USA. 1980; p. 99–114.

[28] Gao D, Mazur P & Critser JK. Fundamental cryobiology of mammalian spermatozoa: scientific principles. In: Karow, A., & Critser, J., editors. Reproductive tissue banking. 1st ed. USA: Academic Press; 1997; p. 263–328.

[29] Mazur P. Freezing of living cells: mechanisms and implications. American Journal of Physiology-Cell Physiology. 1984; 247 (3): C125–C142.

[30] Leibo SP, Semple ME & Kroetsch TG. *In vitro* fertilization of oocytes by 37-year-old cryopreserved bovine spermatozoa. Theriogenology. 1994; 42 (8): 1257–1262.

[31] Horne G, Atkinson AD, Pease EHE, Logue JP, Brison DR & Lieberman BA. Live birth with sperm cryopreserved for 21 years prior to cancer treatment: case report. Human Reproduction. 2004; 19 (6): 1448–1449.

[32] Rofeim O & Gilbert BR. Long-term effects of cryopreservation on human spermatozoa. Fertility and Sterility. 2005; 84 (2): 536–537.

[33] Malik A, Laily M & Zakir MI. Effects of long term storage of semen in liquid nitrogen on the viability, motility and abnormality of frozen thawed Frisian Holstein bull spermatozoa. Asian Pacific Journal of Reproduction. 2015; 4 (1): 22–25.

[34] Córdova A, Strobel P, Vallejo A, Valenzuela P, Ulloa O, Burgos RA, Menarim B, Rodriguez-Gil JE, Ratto MH & Ramírez-Reveco A. Use of hypometabolic TRIS extenders and high cooling rate refrigeration for cryopreservation of stallion sperm: presence and sensitivity of 5′AMP-activated protein kinase (AMPK). Cryobiology. 2014; 69 (3): 473–481.

[35] Ramírez AR, Castro MA, Angulo C, Ramió L, Rivera MM, Torres M, Rigau T, Rodriguez-Gil JE & Concha II. The presence and function of dopamine type 2 receptors in boar sperm: a possible role for dopamine in viability, capacitation, and modulation of sperm motility. Biology of Reproduction. 2009; 80 (4): 753–761.

[36] Larson JL & Miller DJ. Simple histochemical stain for acrosomes on sperm from several species. Molecular Reproduction and Development. 1999; 52 (4): 445–449.

[37] Grove D. Andrological field diagnosis on cattle in the tropics. 1st ed. *Deutsche Gesellschaft fur Technische Zusammenarbeit GmbH.* (GTZ). 1977; p. 286.

[38] Kathiravan P, Kalatharan J, Karthikeya G, Rengarajan K & Kadirvel G. Objective sperm motion analysis to assess dairy bull fertility using computer-aided system—A review. Reproduction in Domestic Animals. 2011; 46 (1): 165–172.

[39] Budworth PR, Amann RP & Hammerstedt RH. A microcomputer-photographic method for evaluation of motility and velocity of bull sperm. Journal of Dairy Science. 1987; 70 (9): 1927–1936.

[40] Budworth PR, Amann RP & Chapman PL. Relationships between computerized measurements of motion of frozen-thawed bull spermatozoa and fertility. Journal of Andrology. 1988; 9 (1): 41–54.

[41] Amann RP. Can the fertility potential of a seminal sample be predicted accurately? Journal of Andrology. 1989; 10: 89–98.

Chapter 7

The Roles of Antioxidants and Fatty Acids in Sperm Cryopreservation

Nurhusien Yimer, Asmatullah Kaka,
Rosnina Yusoff and Abdul Wahid Haron

Additional information is available at the end of the chapter

http://dx.doi.org/10.5772/65571

Abstract

Despite research developments in the area of sperm storage, it has become inevitable to realize a marked reduction in the quality of fresh semen following cryopreservation. As a result, research has continued and will also continue in the future looking forward for a much better and improved methods of sperm cryopreservation along with better understanding of the physical and biochemical challenges that the sperm has to face to survive during freezing. Among the various attempts made to improve the cryopreservation process and subsequently result in superior quality of sperm after thawing include manipulating the composition of semen extenders by addition of exogenous products including antioxidants and fatty acids. While fatty acids are added to strengthening plasma membrane stability, Antioxidants are incorporated to compensate the reduction in the endogenous antioxidants level of seminal plasma due to dilution as well as to combat with the excess reactive oxygen species (ROS) production during freezing. In this chapter, the roles of antioxidants and fatty acids in mammalian sperm cryopreservation, both from endogenous and exogenous perspectives, will be discussed with reference to the latest research findings.

Keywords: antioxidants, fatty acids, sperm cryopreservation

1. Introduction

Demand for livestock products are increasing throughout the world, due to the shifting of consumption patterns toward livestock products and increase in human population. For example, meat and dairy consumption over the last decade increased at a rate of 3–5% annually in Asian countries [1]. To fulfill the increasing demand for animal products, there is a need to increase animal production by improved reproductive technologies. One of the reproductive technologies that have tremendously contributed to the genetic improvement and development of animal production especially, in the dairy cattle industry is artificial insemination (AI).

AI is a process of depositing sperm manually into a female reproductive tract (usually, uterus, or cervix) for the purpose of achieving viable pregnancy through *in vivo* fertilization using a method other than natural mating. The first documented history of successful use of AI backdates to 1780 by L. Spallanzani who experimented in a bitch that subsequently gave birth to three pups. Later in about 1900, research on AI continued in farm animals and subsequently E.I. Ivanoff who initially studied in horses became the first to successfully inseminate cattle and sheep [2]. Since then, AI has undergone tremendous advances in techniques and applications in a wide variety of species of animals and human.

AI provides a lot of advantages over natural breeding. It maximizes the lifespan reproductive potential of a given male as a single semen ejaculate can be diluted and used to inseminate several females. Other prominent advantages of using AI in farm animals include improvement of genetics through more accurate evaluation of breeder males and greater use of superior germplasm, control of sexually transmitted diseases, improved record keeping, and it avoids the cost and necessity of keeping breeder males in the farm. Although the need to detect females on estrus is one of the prominent disadvantages of AI considered, with the development and advances in other assisted reproductive techniques (ART) such as estrus synchronization and timed AI as well as heat detection aids, this disadvantage of AI is dwindling.

AI can be performed using either fresh or cryopreserved (frozen-thawed) semen. Although the use of fresh semen in AI results in a higher success rate than using cryopreserved semen, it requires keeping males for semen collection in nearby place and immediate shipment of the semen for insemination; otherwise, the semen quality will quickly deteriorate. However,

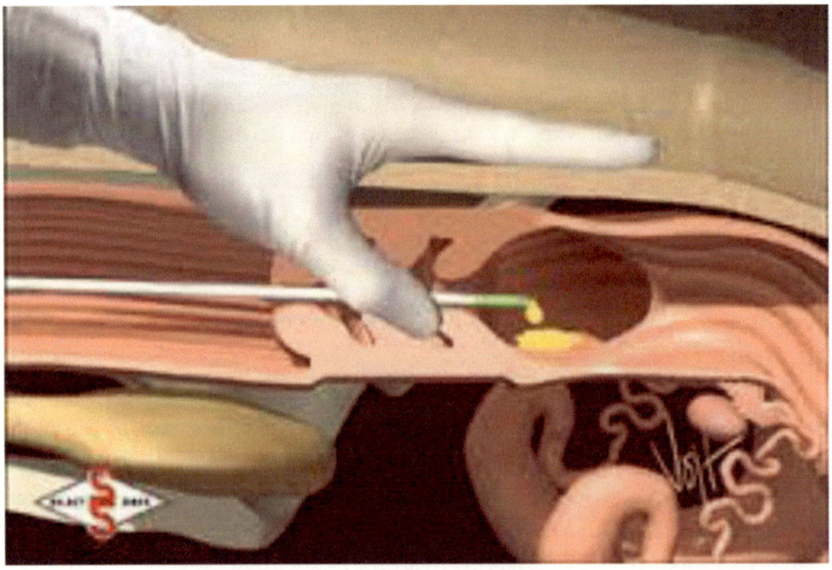

Figure 1. Proper placement of insemination gun to deposit semen in the body of the uterus [61].

http://dx.doi.org/10.5772/65571

cryopreservation provides the opportunity to store the semen for a longer period of time as well as an easy global shipment of superior germplasm to be used for AI. According to Turner [3], it has been documented that AI was estimated to be used on over 110 million breeding cattle (including buffaloes) globally by the turn of the twenty-first century, with the largest number of AIs (50 million) being carried out in the Far East (mainly China, India and Pakistan). There were 1600 cattle semen banks globally and the cattle industry produced over 260 million doses of bull semen from a relatively small number of 41,000 bulls at 648 collection centers world-wide. There was an international trade in bull semen (most of it Holstein), with over 19 million doses recorded as exported globally. The proportion of all cattle inseminations that were carried out by AI rather than by natural mating accounted for about 61% in Europe and about 25% in North America and the Far East. These reflect the significant contribution of semen cryopreservation to the global application of AI in cattle for genetic merit as well as the impact of the quality of postcryopreserved semen that could have on the successful outcome of AI (**Figure 1**).

2. Semen cryopreservation

Semen cryopreservation is a reproductive biotechnology used to preserve and store sperm at a low-freezing temperature for a short or long period of time for various purposes such as in assisted reproduction technologies (ART), species or breed conservation and fertility treatment as in clinical medicine. As discussed earlier, this technique has played a significant role in the livestock industry by overcoming space, distance, and time limitations for the transport of genetically valuable sperm globally and use of AI. Sperm cryopreservation is also an integral component of human reproductive medicine, recognized as an efficient procedure for management of male fertility before therapy for malignant diseases, vasectomy, or surgical infertility treatments, to store donor and partner spermatozoa before assisted reproduction treatments and to ensure the recovery of a small number of spermatozoa in severe male factor infertility [4].

The observation made by Spallanzani in 1803 stated that sperm cooled with snow was not killed but rendered motionless until exposed to heat, after which they became motile for several hours, and the successful cattle insemination in the 1900s could be considered the initial triggers toward the discovery of the cryopreservation procedure. It was in 1940, A. S. Parkes and C. Polge developed a successful method for sperm freezing and storage at low temperatures (-79°C) using dry ice [2]. The same researchers were able to identify glycerol, which is commonly used as a cryoprotectant up to now, as an important factor that helps to protect fowl sperm during the freezing and thawing process. Later in 1957, the freezing mechanism transformed from the use of dry ice to the use of liquid nitrogen contained in a large stainless steel or aluminum vacuum containers, pioneered by the American Breeders Association [2]. Packaging sperm using plastic tubes (straws) or glass ampules were also important adjunct discoveries in the history of development of successful cryopreservation protocol. From the onset of using frozen semen until about the 1970s, glass ampoules were used almost exclusively for packaging, while the straws (0.5 ml—medium straw; 0.25 ml—mini straw) have been the package of choice from the 1970 up to now. As shown in **Figure 2**, liquid nitrogen tank is used to freeze and store the sperm.

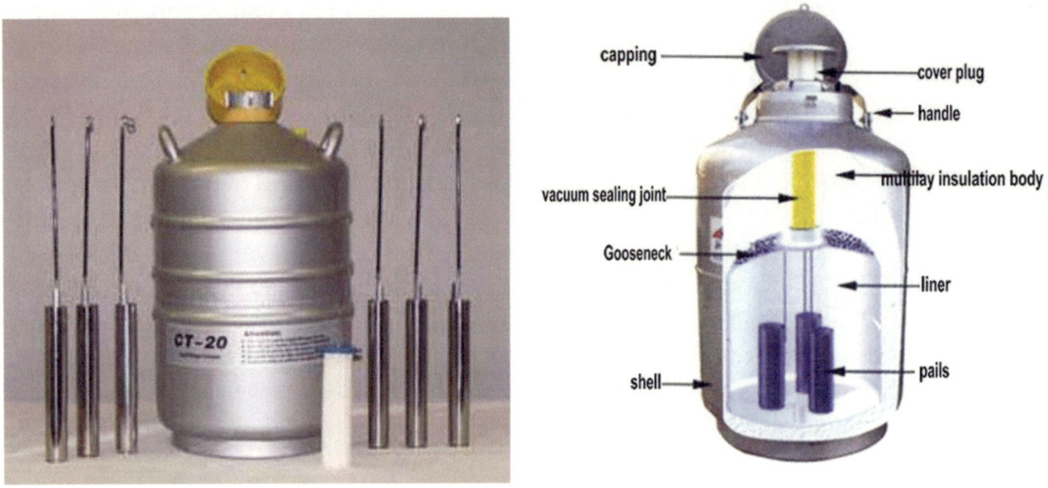

Figure 2. Liquid nitrogen tank used to freeze and store semen in straws (left) and a cross section of a typical semen storage unit (right).

2.1. Semen cryopreservation media

A cryopreservation media, which is also known as semen diluent or extender, plays a crucial role in the quality of sperm after thawing and consequently, it affects significantly the success of AI in the livestock industry. This is because, the survival of the sperm after going through all the physical and biochemical challenges during cooling, freezing, and thawing is dependent mainly on the protective roles provided by the different components of the semen extender used. There are a number of ready-to-use commercially available semen extenders for the production of animals, which include Bioxcel®. Semen extenders such as tris-egg-yolk can also be prepared freshly in the laboratory provided that the components to be mixed are available. In spite of some variations among extenders in their composition, the ultimate objectives intended to be achieved are usually common. These objectives, which have been described as properties that a good semen extender should have [2], include:

1. To have the same free ion concentration or to be isotonic with semen.
2. To have a buffering capacity to maintain pH by neutralizing acid produced by sperm metabolism.
3. To protect the sperm from cold shock injury during cooling from body temperature down to 5°C.
4. Provide nutrition for the sperm metabolism.
5. Control microbial contaminants.
6. Provide sufficient protection to the sperm from damage during freezing and thawing.
7. Preserve the life of the sperm with minimum drop in fertility.

Despite vast research that has been conducted in the area of semen extender's composition and sperm cryopreservation, and remarkable improvements, yet it has remained inevitable for a big portion of the fresh semen to be damaged during cryopreservation and subsequently resulting in a significant decrease in its quality. This might conclude that the above-mentioned properties of a good semen extender are not good enough and the presence of many other unknown damaging factors that need to be identified and addressed as well. Among the other challenges that the sperm has to face during cryopreservation that appears to be overlooked is oxidative stress (OS). Traditional semen extenders lack specific components targeted to deal with oxidative stress or protect sperm from oxidative damage. As a result, recent studies have focused on finding supplements to semen extenders to protect sperm from oxidative stress as well as other supplements that would help to protect sperm damage in structure and function. In this chapter, we focus on antioxidants and fatty acids as supplements for sperm cryopreservation.

3. Oxidative stress and antioxidants

3.1. Oxidative stress (OS)

Before discussing about antioxidants and their role as supplements to protect sperm from oxidative damage, it is also important to review on what oxidative stress mean and evidence that shows sperm quality is indeed affected by oxidative stress. Oxidative stress (OS) refers to a disturbance in the balance between the production of reactive oxygen species (ROS) (free radicals) and the antioxidant defense that helps to counteract or detoxify their harmful effects. As discussed earlier, cryopreservation of sperm is a routine practice especially in cattle breeding industries for the purpose of artificial insemination. The freezing-thawing procedures of cryopreservation are known to produce ROS in sperm samples. Exposure of semen to cold shock and atmospheric oxygen during cryopreservation increases the susceptibility to lipid peroxidation (LPO) due to higher production of ROS [5, 6].

Oxidants, such as reactive oxygen species (ROS) which are produced physiologically in living cells during respiration as well as by abnormal or dead sperm and phagocytic cells of both the ejaculate and female reproductive tract, affect the quality of postthawed sperm in animals [7, 8]. These ROS can inhibit sperm motility, capacitation, and acrosome reaction mediated by lipid peroxidation (LPO) of sperm membrane. Lipid peroxidation has been correlated with exposure of spermatozoa to ROS and it has been demonstrated that spermatozoa undergoing freeze-thaw cycles produces ROS [9]. A lipid peroxidation rate can be assessed by measuring malondialdehyde (MDA) level in the semen [10] which is one of the final products of polyunsaturated fatty acids (PUFAS) peroxidation in the cell and considered to be an oxidative stress biomarker.

Oxidative stress, which occurs when oxidants outnumber antioxidants in tissues or cells causing pathological effects, is known to play a significant role in the pathophysiology of infertility in human [11]. Factors causing oxidative stress such as ROS are known to be involved in multiple physiological processes from oocyte maturation to fertilization, embryo development, and pregnancy [11]. According to Agarwal et al. [12], OS is also considered to be one of the key causes of defective gametes and non- or poorly developing embryos in assisted reproductive techniques (ART). A poor fertilization rate, impaired embryo development, and higher rates of pregnancy loss associated with increasing OS in male germ cells are among the adverse effects recorded [13]. A number of other studies have also confirmed significant pathological effects of OS on gametes, embryos, and subsequent implantation resulting in poor pregnancy outcomes. Sperm DNA damage [14] implicated as the cause of increased incidence of abortion [13], loss of plasma membrane fluidity that leads to decrease in vigor and ultimate immobilization, and decrease in mitochondrial potential that leads to apoptosis are among the pathological effects of OS on sperm reported [15]. While reduced quality, early developmental block, and retardation of embryos [11], high fragmentation, and lower blastulation rate that lead to a lower pregnancy rate [2] are among the pathological effects of OS reported on embryos. Considering these adverse effects of OS on reproduction, ameliorating strategies such as *in vivo* and *in vitro* supplementation of antioxidants have been suggested and implemented with improved results [15, 16].

3.2. Antioxidants and their role in sperm cryopreservation

Antioxidants can be classified as enzymatic and nonenzymatic antioxidants. Enzymatic antioxidants are also known as natural antioxidants; they neutralize excess ROS and prevent it from damaging the cellular structure. Enzymatic antioxidants are composed of superoxide dismutase (SOD), catalase, glutathione peroxidase (GPx), and glutathione reductase (GR), which also cause reduction of hydrogen peroxides to water and alcohol [6, 11]. Nonenzymatic antioxidants are also known as synthetic antioxidants or dietary supplements. The antioxidant system in the body is influenced by dietary intake of antioxidants, vitamins, and minerals such as vitamin C, vitamin E, zinc, taurine, hypotaurine, and glutathione [6, 11].

In mammals, seminal plasma contains a number of antioxidants which include superoxide dismutase, catalase, glutathione peroxidase, free radical scavengers such as vitamins C and E, hypotaurine, taurine, and albumin [17]. The presence of these antioxidants in semen helps to counteract with the oxidants and protect the spermatozoa from damage. Semen dilution using extenders for the purpose of having more doses from a single ejaculate and cryopreservation, however, decreases the concentrations of natural components of antioxidants. The decrease in the components of antioxidants due to dilution coupled with an increase in production of ROS during cryopreservation exacerbates the condition of spermatozoa and further degrades its postthaw quality and fertilizing ability. To minimize the effect of oxidants on a diluted semen, researchers have tested the impact of adding antioxidants into extenders in many species of animals including bull and have observed improvement in the quality of postthaw spermatozoa compared with controls based on conventional andrological tests [8, 17].

There are numbers of antioxidants tested as supplements to mammalian sperm cryopreservation, but perhaps the most frequently studied antioxidant is alpha-tocopherol form of vitamin E. Vitamin E is a fat soluble vitamin that may directly quench the free radicals such as peroxyl and alkoxy (ROO•) generated during ferrous ascorbate-induced LPO; thus, it is suggested as major chain breaking antioxidant and a protectant of LPO and polyunsaturated fatty acids

(PUFAS) in cell membranes from oxidation [6]. Addition of natural antioxidants such as alpha-tocopherol and ascorbate has been reported to have protective effect on metabolic and cellular viability of cryopreserved bovine spermatozoa [18, 19]. More recently, butylated hydroxyl toluene (BHT), a synthetic analogue of vitamin E, has been tested for its antioxidant effect on bulls [8] and buffalo bull spermatozoa [17]. These studies investigated the impact of adding BHT into semen extenders on postthaw semen quality based on conventional andrological tests such as motility and viability. Results of their study indicated improvement of semen quality following cryopreservation compared to untreated controls. A concentration range of BHT added between 0.5 and 1 mM for bull spermatozoa [8] and between 1 and 2 mM for buffalo sperm [17] was reported to be optimum for cryopreservation and better postthaw sperm quality. BHT is a synthetic analogue of vitamin E that controls the auto-oxidation reaction by converting peroxy radicals to hydroperoxides [17]. BHT has also been successfully tested to preserve liquid semen in other species of animals, such as turkey, to minimize cold shock damage in ram, boar, and goat spermatozoa [17]. Recently, BHT has also been tested in our laboratory as a supplement to both lecithin-based and egg-yolk-based extenders for bull sperm cryopreservation [10]. Findings showed that supplementation of BHT improved general motility, progressive motility, morphology, acrosome integrity, DNA integrity, and oxidative stress level of sperm at 0.5 mM/ml for lecithin-based Bioxcel® and at 1–1.5 mM/ml of BHT for tris- and citrate-egg-yolk extenders compared with their controls. However, higher concentrations of 2.0 and 3.0 mM/ml of BHT had a detrimental effect compared with the control of all extenders evaluated and it was concluded that BHT supplementation at lesser concentrations (0.5–1.5 mM/ml) could improve frozen-thawed bull sperm quality by reducing oxidative stress produced during the freezing-thawing procedures in either lecithin or egg-yolk-based extenders.

4. Fatty acids and their role in sperm cryopreservation

Fatty acid is the composite of a hydrocarbon chain, a methyl group and a carboxylic acid group. The hydrocarbon chains vary in length between 14 and 24 carbon atoms and have difference in the number and position of carbon-carbon double bonds. Saturated fatty acids (SFA) contain no double bonds as all carbon atoms are saturated with hydrogen atoms. This gives the general formula of (CH_3 (CH_2) n COOH) where n is the number of hydrocarbon chain [20]. Fatty acids are classified as essential fatty acids (EFA) and nonessential fatty acids (*n*EFA) in mammals. Essential fatty acids cannot be synthesized by the body and must be included in the diet due to a lack of desaturase enzymes and the inability to synthesize fatty acids [21].

Fatty acids are vital components of phospholipids and diglyceride and triglycerides, where they are attached to a glycerol molecule, with an additional polar, organic molecule adjoined to the glycerol molecule of the phospholipid. Phospholipids and diglycerides contain two fatty acid chains while triglycerides contain three. When these fatty acids are not part of a larger structure, they are known as free fatty acids.

Fatty acids contain hydrophobic tail of long carbon chains and a hydrophilic head containing a negatively charged phosphate group [22]. The roles of phospholipids are primarily as constituents of biological membranes [23].

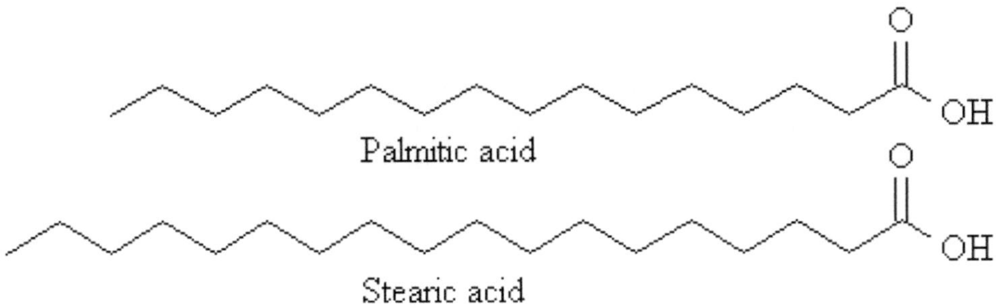

Figure 3. Structure of palmitic acid (upper) and stearic acid (lower). Both have COOH groups to the end with methyl (CH_3) end to the left, C and H atoms not marked.

4.1. Types of fatty acids

Fatty acids are broadly classified as saturated and unsaturated fatty acids.

4.1.1. Saturated fatty acids (SFAS)

Saturated fatty acids (SFAs) are considered as they do not have double bonds in their hydrocarbon chain. The most common SFAs in sperm are palmitic acid (16:0) and stearic acid (SA; 18:0) [24, 25] (**Figure 3**). Because of the absence of double and triple bonds, saturated fatty acids are crowded tighter in cell membranes, therefore, reduce membrane fluidity. Membrane fluidity increases as levels of membrane unsaturation increases [26].

4.1.2. Unsaturated fatty acids

Unsaturated fatty acids are further divided as monounsaturated fatty acids (MUFA) or polyunsaturated (PUFAs). Monounsaturated fatty acid (MUFA) components contain only one double bond and PUFAs contain two or more double bonds. MUFAs and PUFAs are then further classified into three families, omega 3, 6, and 9 (*n*-3, *n*-6, and *n*-9) unsaturated fatty acids according to the distance of the first double bond from the methyl terminal [26].

4.1.2.1. Omega 3, 6 and 9 fatty acids

Omega 3 fatty acid is a group of fatty acids in which the first double bond is located on the third carbon-carbon bond from the methyl end of the hydrocarbon chain. The first double carbon-carbon bond of omega 6 fatty acid is located on the sixth carbon-carbon bond from the methyl end. Omega 9 fatty acids have the first double bond on the ninth carbon-carbon bond from the methyl end group. Omega 3, 6, and 9 fatty acids can also be denoted as *n*-3, *n*-6, and *n*-9 or ω-3, ω-6, and ω-9 fatty acids, respectively. Omega 3 fatty acids in many species of sperm include ALA (18:3), DHA (22:6), DPA (22:5), and eicosapentaenoic acid (EPA; 20:5) (**Figure 4**). Arachidonic acid (AA; 20:4) and LA (18:2) are two *n*-6 fatty acids in sperm (**Figure 5**) while oleic acid (OA; 18:1) is *n*-9 fatty acid family (**Figure 6**) [27].

Animals, such as bull and boar, and humans cannot manufacture fatty acids with carbon chains more than 18 carbons, because of deficiency in the desaturase enzymes at

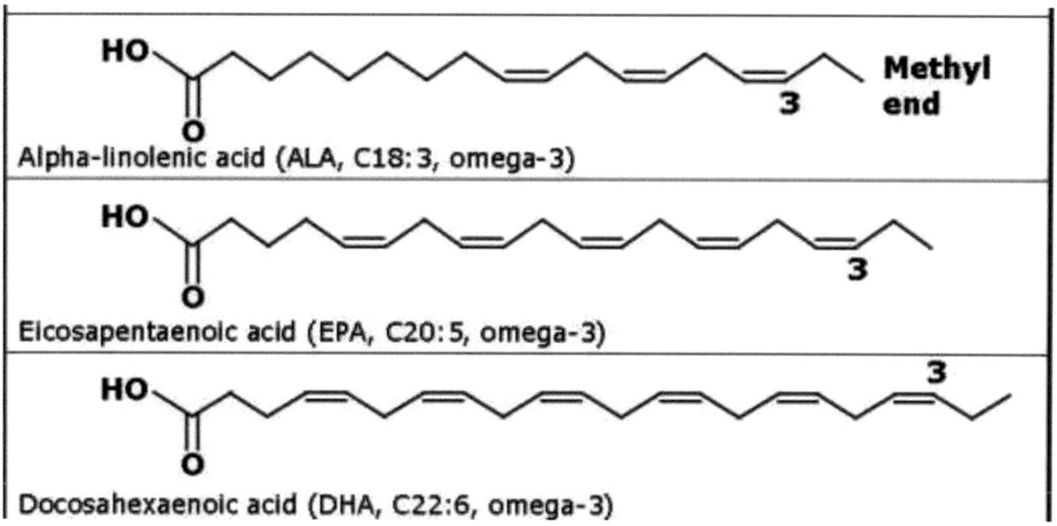

Figure 4. Structure of *n*-3 polyunsaturated fatty acids (PUFAs), alpha-linolenic acid, docosapentaenoic acid, docosahexaenoic acid and eicosapentaenoic acid.

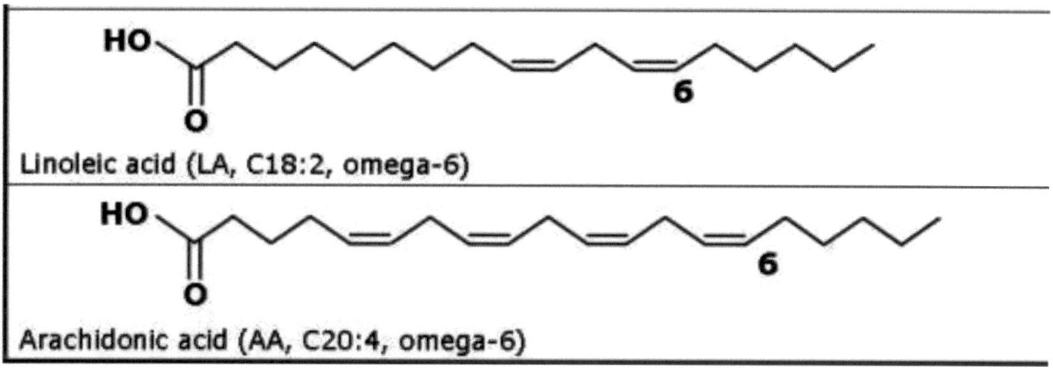

Figure 5. Structure of *n*-6 polyunsaturated fatty acids (PUFAs), arachidonic acid and linoleic acid.

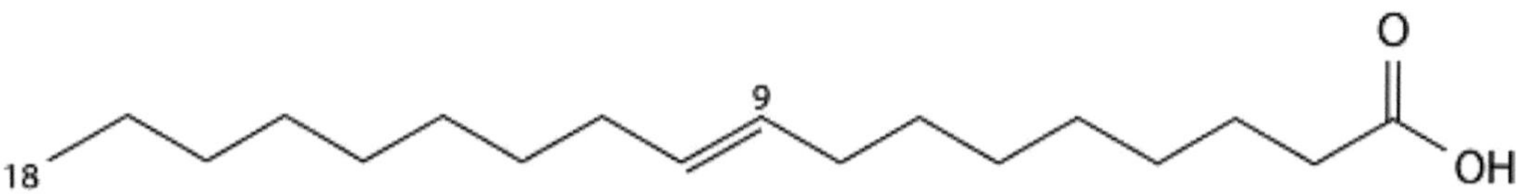

Figure 6. Structure of *n*-9 monounsaturated fatty acids (MUFA) and oleic acid.

Δ1-desaturase, Δ2-desaturase, and Δ3-desaturase enzymes, which could form ALA, LA, and OA, whereas these animals contain Δ4-desaturase, Δ5-desaturase, Δ6-desaturase, and Δ9-desaturase, the number indicating the location that the desaturase enzyme places the double bond in the carbon chain [27]. *n*-3, *n*-6, and *n*-9 fatty acids are commonly known

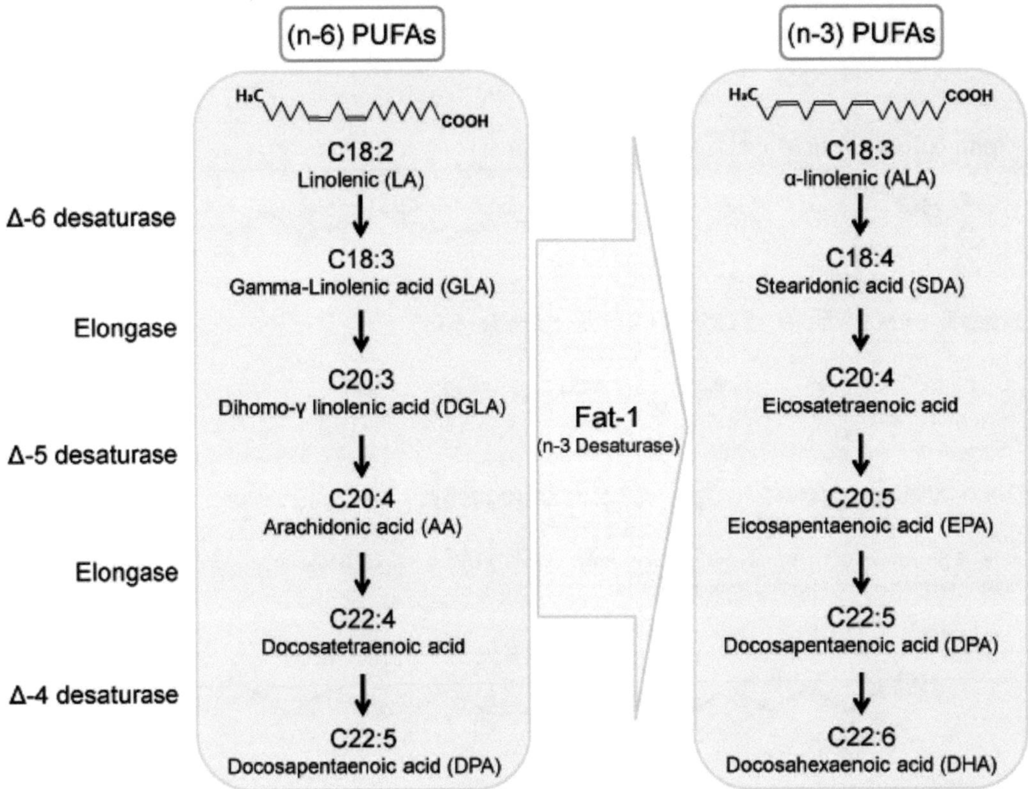

Figure 7. Metabolism of parent fatty acids ALA (*n*-3) and LA (*n*-6) into longer carbon chain fatty acids with relevant enzymatic reactions to form the fatty acids [26].

as the parent fatty acids. These fatty acids are usually found in the diet. Longer chain fatty acids are manufactured by the process of elongation and desaturation reactions generally named as *de novo* synthesis of fatty acids [26, 27] (**Figure 7**).

4.2. Fatty acid compositions of sperm

Epididymis is the store house of sperm where sperm undergo the process of maturation and remodeling of the plasma membrane also occurs. During remodeling, secreted epididymal glycoproteins uptake takes place, consumption of phospholipids from the membrane bilayer and relocation of protein and lipid constituents are restructured during maturation [28] and sperm acquire progressive motility and the ability to fertilize an oocyte [29]. Particularly, bull sperm lose half of their phospholipid and major phospholipid [30]. Fatty acids, as a major component of phospholipids, also undergo a major reduction during epididymal stage [30]. Retention of fatty acids (SFA, MUFAs, and PUFAs) is an indication of immature and defective sperm [25, 31] who studied bull sperm heads and tails described that sperm tails retained more *n*-3 PUFAs than the sperm head while *n*-6 PUFAs were more concentrated in sperm heads than in the tails. Same patterns of *n*-3 and *n*-6 by were found in human sperm. A higher

Type	Name	Carbon chain length
Saturated fatty acids	Myristic	C14:0
	Palmitic	C16:0
	Palmitoleic	C16:1
	Stearic	C18:0
Unsaturated fatty acids	***n*-9 PUFAs;**	
	Oleic acid	C18:1*n*-9OA
	n-6 PUFAs	
	Linoleic acid	C18:2*n*-6
	Gamma linolenic acid (GLA)	C18:3*n*-6
	Arachidonic acid (AA)	C20:4*n*-6
	***n*-3 PUFAs**	
	Eicosapentaenoic acid (EPA)	C20:5*n*-3
	Alpha-Linolenic acid (ALA)	C18:3*n*-3
	Docosapentaenoic acid (DPA)	C22:5*n*-3
	Docosahexaenoic acid (DHA)	C22:6*n*-3

Table 1. Fatty acids found in sperm of different animals [35].

percentage of *n*-6 fatty acids (28%) were found in the total bull sperm than *n*-3 fatty acids (23%) in both bull and human sperm. Poulos et al. [30] reported that DHA was one of the main fatty acids of caudal and ejaculated bull sperm and human sperm, respectively. Lenzi et al. [26] has suggested that up to 60% of PUFA in normal human sperm consists of DHA; however, Zalata et al. [32] also reported the same.

Palmitic acid and stearic acid have been identified as the most saturated fatty acids of whole human sperm [25, 26]. Human sperm fatty acid from asthenozoospermic (low motility and viability) males differed from normospermic (normal) males in composition. The former showed lower of DHA but higher OA levels [32]. While unsaturated fatty acids, as a whole, were reduced in the asthenozoospermic males compared to normospermic males [33]. However, infertile human males were found to have higher levels of *n*-6 PUFAs, which decreased sperm concentration, decreased motility, and higher abnormal count [34]. High levels of *n*-3 PUFAs (ALA, DHA, DPA, and EPA) were linked with sperm development, improved motility, and morphology and cryogenic resistance [34]. The superabundance of unsaturated fatty acids, leave sperm extremely susceptible to reactive oxygen species (ROS) attack, oxidative stress (OS), and lipid peroxidation (LPO) [24, 25] (**Table 1**).

4.3. Roles of fatty acids in sperm cryopreservation

Adenosine triphosphate (ATP) produces anaerobic and aerobic respiration and provides energy within the cells for sperm functions. ATP produced through glycolysis (anaerobic respiration) is a major source of ATP in sperm. Glycolysis occurs in the cytosol of sperm; hence it distributes ATP uniformly in sperm. Mitochondria present in sperm midpiece produce 15 times more ATP by oxidative phosphorylation or aerobic respiration. The aerobic respiration requires oxygen (O_2) and carried out through electron transport chain (ETC) is an effective energy production method. ROS is the by-product of the ETC by leaking of electron and formation of superoxide during respiration [36]. The mitochondrial ETC is composed of four (Complex I–IV) multiprotein complexes and many electron carriers, i.e., flavoproteins, iron-sulfur proteins, ubiquinone, and cytochromes [37]. Electrons are uptaken on ETC by complex I and complex II. Complex I carries electron from nicotinamide adenine dinucleotide (NADH) and nicotinamide and complex II from succinate ($C_4H_6O_4$) [37]. Succinate is a flavin adenine dinucleotide (FAD) linked substrate, which acts as a coenzyme in redox reactions in the body, and later electrons move through carrier to complex III by coenzyme Q (CoQ) or ubiquinone and after that by cytochrome C transport electrons to complex IV [37]. During transport at ETC, electrons escape and form superoxide, which are then transformed to ROS, i.e., hydroperoxyl, hydrogen peroxide, and hydroxide radicals Superoxide are formed at complex I and complex III and O_2 is fully reduced to water (H_2O) at the end of the ETC [37].

Nicotinamide adenine dinucleotide (NADH) and nicotinamide adenine dinucleotide phosphate (NADPH) are reducing agents and provide protons within a cell. NADPH and NADH also are the source of increased ROS by supplying electrons for the formation of free radicals, by the reduction of oxygen to superoxide [38]. NADPH is present at the residual cytoplasmic droplet and triggers a NADPH oxidase (NOX) system in the human sperm plasma membrane [39]. According to the previous studies, human sperm generate ROS using the NOX5 enzyme with increase in calcium ions (Ca^{2+}) [40]. NADPH oxidases are plasma enzymes that catalyze the production of ROS by electron flow from NADPH to surrounding cell membrane to molecular oxygen, in order to form superoxide by reduction [41].

The supplementation of exogenous NADPH human sperm encouraged superoxide generation [39] and exogenous NADPH stimulate ROS effectively, necessarily, is to penetrate the sperm membrane and results that sperm damaged membranes showed a higher tendency to absorb NADPH and as a consequence form ROS. NADPH production in the cytoplasm is named as the monophosphate shunt, regulated by enzyme glucose-6-phosphate dehydrogenase. This enzyme also controls the glucose efflux rate and the presence of this enzyme itself is an indicator for immature human sperm [38].

Many studies have been conducted to determine effect of fatty acids particularly polyunsaturated fatty acids on cooled, chilled, and frozen-thawed semen quality in different species of animals. Some of the testimonies are discussed below.

4.3.1. Bulls

Kiernan et al. [42] determined that ALA maintained sperm motility at 100 μM and viability at 10 and 50 μM in citrate-based extender in bull semen chilled for 7 days. Palmitic acid and oleic acid maintained motility and viability at 50 and 100 μM. Takahashi et al. [43] reported that addition of linoleic acid improved frozen-thawed spermatozoa motility and viability of bull semen. Nasiri et al. [44, 45] added DHA improved sperm quality of frozen-thawed quality of bull sperm. Dietary ALA improved the plasma membrane integrity, acrosome integrity, and DNA integrity of frozen-thawed spermatozoa [46]. In feed also resulted in improved motility in fresh semen of bull [47, 48]. Kaka et al. [35, 49, 50] reported that individual addition of ALA and DHA in tris and bioxcell extender improved cooled and frozen-thawed quality of bull semen while combination of ALA and DHA decreased semen quality after freezing. Kandelousi et al. [51] and Abavisani et al. [52] also reported that omega-3 PUFAs did not improve motility, progressive, morphology, and viability in citrate extender in frozen-thawed quality of bull semen.

4.3.2. Goat

In vitro addition of omega-3 increased the quality of frozen-thawed spermatozoa in goats [53]. Supplementation of egg-yolk DHA rich in citrate extender also improved the total motility, progressive motility, viability, and morphology of frozen-thawed goat spermatozoa [54].

4.3.3. Sheep

Samadian et al. [55] and Towhidi and Parks [45] tested omega-3 fatty acids and reported the improved frozen-thawed quality of semen in rams.

4.3.4. Buffalo

Fatty acids such as arachidonic acid improved postthawed spermatozoa motility, membrane integrity, acrosome integrity, viability, and DNA of buffalo bull spermatozoa [7].

4.3.5. Human

Omega-3 fatty acid is higher in fertile men than in the infertile so that omega-6 fatty acid is important for sperm quality (Saferinajad et al., 2010). The existence of surplus of unsaturated fatty acids in defective human spermatozoa may increase the oxidative stress which reduces in male fertility [34].

4.3.6. Boar

Boar spermatozoa motility, viability, and acrosome integrity were also improved following addition of linoleic acid, oleic acid, and arachidonic acid [56–58]. Chanapiwat et al. [59] and Kaeoket et al. [60] added that DHA improved motility, membrane integrity, and acrosome integrity, viability, and DNA integrity in boar sperm when used alone and in combination with *L*-cysteine in lactose-egg-yolk extender.

5. Conclusion

As research studies show, it is evident that supplementing semen extenders, with some antioxidants, such as vitamin E and its synthetic analogue, BHT, and fatty acids such as alpha linoleic acid, is beneficial to the sperm to endure the physical and biochemical changes and challenges faced during cryopreservation and results in a superior postthaw quality available for AI. However, it is worth noting that supplements work best at a particular optimum concentration that varies with the type of extender used, as well as the species of the animal from which the sperm has come from. Hence, prior use of a particular antioxidant or fatty acid as a supplement, proper investigation needs to be conducted to determine their optimum concentration to be added into a specific semen extender type.

Author details

Nurhusien Yimer[1]*, Asmatullah Kaka[2], Rosnina Yusoff[1] and Abdul Wahid Haron[1]

*Address all correspondence to: nurdeg2006@gmail.com

1 Theriogenology and Cytogenetics Unit, Department of Veterinary Clinical Studies, Faculty of Veterinary Medicine, Universiti Putra Malaysia, Selangor, Malaysia

2 Department of Animal Reproduction, Faculty of Animal Husbandry and Veterinary Sciences, Sindh Agriculture University Tandojam, Sindh, Pakistan

References

[1] FAO. Statistical Yearbook 2013: *World Food and Agriculture*. Italy: FAO Food Agric Organization UN Rome, Italy. 2013.

[2] Bearden, H. J. and J. W. Fuquay, 1984: *Applied animal reproduction*. Reston: Reston Publishing Company, Inc.

[3] Turner, J., 2010: *Animal breeding, welfare and society*. Routledge, United kingdom.

[4] Di Santo, M., N. Tarozzi, M. Nadalini and A. Borini, 2012: Human sperm cryopreservation: update on techniques, effect on DNA integrity, and implications for ART. *Advances in urology*, 2012, 1–12.

[5] Bucak, M. N., A. Atessahin and A. Yüce, 2008: Effect of antioxidants and oxidative stress parameters on ram semen after the freeze–thawing process. *Small ruminant research,* 75, 128–134.

[6] Bansal, A. K. and G. S. Bilaspuri, 2009: Antioxidant effect of vitamin E on motility, viability, and lipid peroxidation of cattle spermatozoa under oxidative stress. *Animal science papers and reports,* 1, 5–14.

[7] Ejaz, R., M. Ansari, B. Rakha, N. Ullah, A. Husna, R. Iqbal and S. Akhter, 2014: Arachidic acid in extender improves post-thaw parameters of cryopreserved Nili-Ravi buffalo bull semen. *Reproduction in domestic animals,* 49, 122–125.

[8] Asadpour, R. and H. Tayefi-Nasrabadi, 2012: The effect of butylated hydroxytoluene (BHT) on bull spermatozoa frozen in two different extenders. *Comparative clinical pathology,* 21, 577–581.

[9] Alvarez, J.G. and B.T. Storey, 1992: Evidence for increased lipid peroxidative damage and loss of superoxide dismutase activity as a mode of sublethal cryodamage to human sperm during cryopreservation. *Journal of anthology,*13(3), 232–241.

[10] Khumran, A., N. Yimer, Y. Rosnina, M. Ariff, H. Wahid, A. Kaka, M. Ebrahimi and K. Sarsaifi, 2015: Butylated hydroxytoluene can reduce oxidative stress and improve quality of frozen–thawed bull semen processed in lecithin and egg-yolk based extenders. *Animal reproduction science,* 163, 128–134.

[11] Agarwal, A., S. Gupta and R. K. Sharma, 2005: Role of oxidative stress in female reproduction. *Reproductive biology and endocrinology,* 3, 1.

[12] Agarwal, A., G. Virk, C. Ong and S. S. du Plessis, 2014: Effect of oxidative stress on male reproduction. *The world journal of men's health,* 32, 1–17.

[13] Baker, M. A. and R. J. Aitken, 2005: Reactive oxygen species in spermatozoa: methods for monitoring and significance for the origins of genetic disease and infertility. *Reproductive biology and endocrinology,* 3, 67.

[14] Aitken, R. J. and C. Krausz, 2001: Oxidative stress, DNA damage and the Y chromosome. *Reproduction,* 122, 497–506.

[15] du Plessis, S. S., K. Makker, N. R. Desai and A. Agarwal, 2008: Impact of oxidative stress on IVF. *Expert review of obstetrics & gynecology,* 3, 539–554.

[16] Taylor, C. T., 2001: Antioxidants and reactive oxygen species in human fertility. *Environmental toxicology and pharmacology,* 10, 189–198.

[17] Ijaz, A., A. Hussain, M. Aleem, M. Yousaf and H. Rehman, 2009: Butylated hydroxytoluene inclusion in semen extender improves the post-thawed semen quality of Nili-Ravi buffalo (Bubalus bubalis). *Theriogenology,* 71, 1326–1329.

[18] Beconi, M., M. Affranchino, L. Schang and N. Beorlegui, 1991: Influence of antioxidants on SOD activity in bovine sperm. *Biochemistry international,* 23, 545–553.

[19] Beconi, M., C. Francia, N. Mora and M. Affranchino, 1993: Effect of natural antioxidants on frozen bovine semen preservation. *Theriogenology,* 40, 841–851.

[20] Christie, W. W., 1973: *Lipid analysis.* Oxford: Pergamon press Oxford.

[21] Gurr, M. I., J. L. Harwood, and K. N. Frayn, 2008: Chapter 5. Lipid transport. In: *Lipid biochemistry,* 5th edition. DOI: 10.1002/9780470774366.ch5

[22] Vance, D. and J. Vance, 1996: Physical properties and functional roles of lipids in membranes. *Biochemistry of lipids, lipoproteins and membranes,* 31, 1–33.

[23] Hames, B. and N. Hooper, 1997: Instant notes in biochemistry. *Biochemical education,* 25, 253–254.

[24] Aitken, R. J. and M. A. Baker, 2006: Oxidative stress, sperm survival and fertility control. *Molecular and cellular endocrinology,* 250, 66–69.

[25] Koppers, A. J., M. L. Garg and R. J. Aitken, 2010: Stimulation of mitochondrial reactive oxygen species production by unesterified, unsaturated fatty acids in defective human spermatozoa. *Free radical biology and medicine,* 48, 112–119.

[26] Lenzi, A., M. Picardo, L. Gandini and F. Dondero, 1996: Lipids of the sperm plasma membrane from polyunsaturated fatty acids considered as markers of sperm function to possible scavenger therapy. *Human reproduction update,* 2, 246–256.

[27] Gill, I. and R. Valivety, 1997: Polyunsaturated fatty acids, part 1: occurrence, biological activities and applications. *Trends in biotechnology,* 15, 401–409.

[28] Jones, R., 1997: Plasma membrane structure and remodelling during sperm maturation in the epididymis. *Journal of reproduction and fertility. Supplement,* 53, 73–84.

[29] Gatti, J. L., S. Castella, F. Dacheux, H. Ecroyd, S. Metayer, V. Thimon and J. L. Dacheux, 2004: Post-testicular sperm environment and fertility. *Animal reproduction science,* 82, 321–339.

[30] Poulos, A., A. Darin-Bennett and I. White, 1973: The phospholipid-bound fatty acids and aldehydes of mammalian spermatozoa. *Comparative biochemistry and physiology Part B: Comparative biochemistry,* 46, 541–549.

[31] Ahluwalia, B. and R. Holman, 1969: Fatty acid composition of lipids of bull, boar, rabbit, and human semen. *Journal of reproduction and fertility,* 18, 431–437.

[32] Zalata, A. A., A. B. Christophe, C. E. Depuydt, F. Schoonjans and F. H. Comhaire, 1998: The fatty acid composition of phospholipids of spermatozoa from infertile patients. *Molecular human reproduction,* 4, 111–118.

[33] Calamera, J., M. Buffone, M. Ollero, J. Alvarez and G. Doncel, 2003: Superoxide dismutase content and fatty acid composition in subsets of human spermatozoa from normozoospermic, asthenozoospermic, and polyzoospermic semen samples. *Molecular reproduction and development,* 66, 422–430.

[34] Safarinejad, M. R., S. Y. Hosseini, F. Dadkhah and M. A. Asgari, 2010: Relationship of omega-3 and omega-6 fatty acids with semen characteristics, and anti-oxidant status of seminal plasma: a comparison between fertile and infertile men. *Clinical nutrition,* 29, 100–105.

[35] Kaka, A., W. Haron, R. Yusoff, N. Yimer, A. Khumran, K. Sarsaifi, A. A. Behan, U. Kaka, A. A. Memon and M. Ebrahimi, 2015a: Effect of docosahexanoic acid on quality of frozen–thawed bull semen in BioXcell extender. *Reproduction, fertility and development.*

[36] Turrens, J. F., 2003: Mitochondrial formation of reactive oxygen species. *The journal of physiology,* 552, 335–344.

[37] Liu, Y., G. Fiskum and D. Schubert, 2002: Generation of reactive oxygen species by the mitochondrial electron transport chain. *Journal of neurochemistry,* 80, 780–787.

[38] Aitken, J., D. Buckingham and C. Krausz, 1994: Relationships between biochemical markers for residual sperm cytoplasm, reactive oxygen species generation, and the

presence of leukocytes and precursor germ cells in human sperm suspensions. *Molecular reproduction and development*, 39, 268–279.

[39] Aitken, R. J., H. M. Fisher, N. Fulton, E. Gomez, W. Knox, B. Lewis and S. Irvine, 1997: Reactive oxygen species generation by human spermatozoa is induced by exogenous NADPH and inhibited by the flavoprotein inhibitors diphenylene iodonium and quinacrine. *Molecular reproduction and development*, 47, 468–482.

[40] Bánfi, B., G. Molnár, A. Maturana, K. Steger, B. Hegedûs, N. Demaurex and K. H. Krause, 2001: A Ca2+-activated NADPH oxidase in testis, spleen, and lymph nodes. *Journal of biological chemistry*, 276, 37594–37601.

[41] Fulton, D. J., 2009: Nox5 and the regulation of cellular function. *Antioxidants & redox signaling*, 11, 2443–2452.

[42] Kiernan, M., A. Fahey and S. Fair, 2013: The effect of the in vitro supplementation of exogenous long-chain fatty acids on bovine sperm cell function. *Reproduction, fertility and development*, 25, 947–954.

[43] Takahashi, T., R. Itoh, H. Nishinomiya, M. Katoh and N. Manabe, 2012: Effect of linoleic acid albumin in a dilution solution and long-term equilibration for freezing of bovine spermatozoa with poor freezability. *Reproduction in domestic animals*, 47, 92–97.

[44] Nasiri, A., A. Towhidi and S. Zeinoaldini, 2012: Combined effect of DHA and α-tocopherol supplementation during bull semen cryopreservation on sperm characteristics and fatty acid composition. *Andrologia*, 44, 550–555.

[45] Towhidi, A. and J. Parks, 2012: Effect of n-3 fatty acids and α-tocopherol on post-thaw parameters and fatty acid composition of bovine sperm. *Journal of assisted reproduction and genetics*, 29, 1051–1056.

[46] Calisici, O. 2010: Investigation of antioxidative capacity in bovine seminal plasma–effects of omega-3 fatty acids. *Doctor of veterinary medicine (thesis). University of Veterinary Medicine, Hanover, Germany.*

[47] Kelso, K., A. Redpath, R. Noble and B. Speake, 1997: Lipid and antioxidant changes in spermatozoa and seminal plasma throughout the reproductive period of bulls. *Journal of reproduction and fertility*, 109, 1–6.

[48] Gholami, H., M. Chamani, A. Towhidi and M. Fazeli, 2010: Effect of feeding a docosahexaenoic acid-enriched nutriceutical on the quality of fresh and frozen-thawed semen in Holstein bulls. *Theriogenology*, 74, 1548–1558.

[49] Kaka, A., H. Wahid, Y. Rosnina, N. Yimer, A. Khumran, A. A. Behan and M. Ebrahimi, 2015b: Alpha-linolenic acid supplementation in tris extender can improve frozen–thawed bull semen quality. *Reproduction in domestic animals*, 50, 29–33.

[50] Kaka, A., H. Wahid, Y. Rosnina, N. Yimer, A. Khumran, K. Sarsaifi, A. A. Behan, U. Kaka and M. Ebrahimi, 2015c: α-Linolenic acid supplementation in BioXcell® extender can

improve the quality of post-cooling and frozen-thawed bovine sperm. *Animal reproduction science,* 153, 1–7.

[51] Kandelousi, M. S., J. Arshami, A. Naserian and A. Abavisani, 2016: The effects of addition of omega-3, 6, 9 fatty acids on the quality of bovine chilled and frozen-thawed sperm. *Open veterinary journal,* 3, 47–52.

[52] Abavisani, A., J. Arshami, A. A. Naserian, M. A. S. Kandelousi and M. Azizzadeh, 2013: Quality of bovine chilled or frozen-thawed semen after addition of omega-3 fatty acids supplementation to extender. *International journal of fertility & sterility,* 7, 161.

[53] Ansari, M., A. Towhidi, M. Moradi Shahrbabak and M. Bahreini, 2012: Docosahexaenoic acid and alpha-tocopherol improve sperm cryosurvival in goat. *Slovak journal of animal science,* 45, 7–13.

[54] Yimer, N., Y. Rosnina, H. Wahid, A. Saharee, K. Yap, P. Ganesamurthi and M. Fahmi, 2011: Trans-scrotal ultrasonography and breeding soundness evaluation of bulls in a herd of dairy and beef cattle with poor reproductive performance. *Pertanika journal of tropical agricultural science,* 34, 217–228.

[55] Samadian, F., A. Towhidi, K. Rezayazdi and M. Bahreini, 2010: Effects of dietary n-3 fatty acids on characteristics and lipid composition of ovine sperm. *Animal,* 4, 2017–2022.

[56] Hossain, M. S., K. Tareq, K. I. Hammano and H. Tsujii, 2007: Effect of fatty acids on boar sperm motility, viability and acrosome reaction. *Reproductive medicine and biology,* 6, 235–239.

[57] Rooke, J., C. Shao and B. Speake, 2001: Effects of feeding tuna oil on the lipid composition of pig spermatozoa and in vitro characteristics of semen. *Reproduction,* 121, 315–322.

[58] Castellano, C. A., I. Audet, J. Bailey, J. P. Laforest and J. Matte, 2010: Dietary omega-3 fatty acids (fish oils) have limited effects on boar semen stored at 17 C or cryopreserved. *Theriogenology,* 74, 1482–1490.

[59] Chanapiwat, P., K. Kaeoket and P. Tummaruk, 2012: Improvement of the frozen boar semen quality by docosahexaenoic acid (DHA) and L-cysteine supplementation. *African journal of biotechnology,* 11, 3697–3703.

[60] Kaeoket, K., P. Sang-urai, A. Thamniyom, P. Chanapiwat and M. Techakumphu, 2010: Effect of docosahexaenoic acid on quality of cryopreserved boar semen in different breeds. *Reproduction in domestic animals,* 45, 458–463.

[61] Selk, G. 2002: Artificial Insemination for Beef Cattle. Retrieved from http://www.thebeefsite.com/articles/721/artificial-insemination-for-beef-cattle/

Chapter 8

Cryopreservation of Epididymal Sperm

Melina A.F. Bertol

Additional information is available at the end of the chapter

http://dx.doi.org/10.5772/65010

Abstract

The recovery and cryopreservation of epididymal sperm are essential tools to preserve genetic stocks of valuable domestic or wild animals under adverse conditions and also as an alternative source of gametes in cases of human infertility. This technique is recommended after death or when it is not possible to recover semen by usual methods. Sperm from the epididymis has been studied by many authors in different species. Cryopreservation is the most effective method for long-term preservation of genetic material from valuable breeding individuals. The protocols and diluents used for cryopreservation of conventional semen are well established, but many challenges remain working with sperm extracted directly from the epididymis. The spermatozoa retrieved from the cauda epididymis have special features, such as the absence of seminal plasma and large numbers of distal cytoplasmic droplets, which necessitate special handling, both for cryopreservation and for fertilization. For these reasons, it is important to describe in detail the features needed to cryopreserve sperm from the epididymis.

Keywords: cryopreservation, epididymis, fertility, reproduction, sperm

1. Introduction

The cryopreservation technique is widely used in reproductive biology because it facilitates the application of assisted reproduction techniques such as artificial insemination and *in vitro* fertilization (IVF). Besides that it allows long-term storage of gametes, facilitating transport over long distances and the formation of a genetic bank, essential for the conservation of species. It is an important complementary tool for conservation to avoid excessive losses of genetic variation or races of extinction. Cryopreservation is considered to be the most effective way for long-term preservation of genetic materials, cells, tissues and microorganisms. It allows to maintain biological material at very low temperatures, indefinitely. The expansion of genetic material from high-value breeders became possible with the advent of semen cryopreservation. The main advantages of the technique are: portability for world trade because it is not necessary to purchase or move the males, and it also acts in the prevention of diseases capable of transmission through natural breeding.

Semen constitutes the union of sperm with seminal plasma, secreted by the male accessory glands. Sperms are produced in the testicles (gametogenesis), and maturation occurs in the epididymis when cells enter in the caput of epididymis, progress to the corpus, and finally reach the cauda region, where they are stored until the moment of ejaculation. Spermatozoa stored in the cauda region are generally of good quality and have a high level of maturity, being able to fertilize oocytes. The epididymis provides a favorable environment to retain the sperm with fertilizing capacity for several weeks. Therefore, the cauda epididymis is a major source of gametes of a breeder and in special cases they can be retrieved directly from it.

The sperm recovery from the epididymis is the last chance to use gametes from dead breeders, in addition to enabling the maintenance of a germplasm bank of animals of commercial interest or threatened with extinction [1]. A factor to be considered is the difficulty in collecting semen from wild species and the unexpected death of animals of zoological interest. In this case, the technic can be used to ensure the rescue of male gametes and the preservation by cryopreservation process for maintaining germplasm banks [2]. For animal production as livestock, the sperm recovery directly from the epididymis is a viable alternative when there is a sudden death of breeder of high commercial value, to increase their genetic stock or in cases of bulls unable to ejaculate for some reason [3]. It is the last alternative for future use of gametes from a breeder in assisted reproduction programs. In equine, this technique is important because there is a high incidence of accidental and unexpected death due to high incidence of colic or severe traumatic accidents that compromise the reproductive life of the stallion.

The first important point is the recovery of still viable sperm with good parameters of motility, concentration, and morphological defects as soon as possible after the male death. There is a time limit for this to successfully occur, usually dependent on temperature. The first step is getting the epididymis in the field, after the death of the animal. For handling ease, due to their anatomical location, the testes can be removed with them. Then both should be sent to the laboratory in suitable containers, with or without cooling. In the lab, the trained technicians perform gonads cleaning, epididymis isolation and spermatozoa recovery. Then it can be used in three different ways: first, shortly after harvesting; second, chilled, and third, after cryopreservation process. The last increases the availability time of gametes for application of assisted reproduction techniques. Next we will focus on the key topics relevant to the understanding of the epididymis spermatozoa cryopreservation technique.

2. Epididymal spermatozoa

When the spermatozoon leaves the testicle to the epididymis, it is a non-functional and infertile gamete. Only after the passage through the epididymis, it becomes mature and acquires progressive motility to become able to fertilize the oocyte. This maturation is complex and involves several factors, including the interaction of sperm with proteins that are synthesized in each epithelium region of the epididymis. Sperm passively migrates through the epididymis and after contact with the epididymal secretions get low-molecular weight and water-soluble compounds by an isovolumetric regulatory process. These compounds can be spent together with the cell water when the cells come into contact with hypo-osmotic fluids from accessory glands or genital female tract. The volume adjustment process serves to maintain the volume pattern into the sperm cell and to prevent angulation of the flagellum, which prevents the sperm to migrate efficiently in the female reproductive tract, being unable to fertilize oocytes. The channels responsible for this regulation are located in cytoplasmic droplets. The spermatozoa located in the cranial portion of the epididymis (caput) are considered immature to present an osmolyte content reduced insufficient for complete regulation of cell volume when exposed to hypotonic solutions [4]. The movement of the sperm through the epididymis is performed mainly by the contraction of smooth muscles of the wall of the Caput and corpus region. On the cauda, the smooth muscles of the epididymal duct is generally at rest until it is stimulated to contract at the time of ejaculation for the release of sperm, thus it is responsible for the protection and storage of sperm until ejaculation.

There are two moments prior to recovery of sperm that interferes directly in the success of the process. The first is the time from the death of the animal until the necropsy, and the second is the period of recovery of the gonads to obtain the gametes in the laboratory [5]. The higher the time of the gametes permanence in the cauda epididymis after death or after orchiectomy, the greater the damage to the sperm cell. At room temperature, motility is the first parameter affected. With the increase in the hours before recovery, there is a reduction in the percentage of moving spermatozoa, worsening from 24 hours [6, 7].

There are two moments for the recovery of spermatozoa from the epididymis:

a. After death

b. After orchiectomy

The most common form of spermatozoa recovery from the epididymis is after death. It is indicated in cases of use of last spermatic reserve after breeder death. It is usually recommended when the breeder is found after dead in the field in cases of sudden death from a serious illness, accidents, poisoning and stress problems. It is worth mentioning that infectious diseases can contaminate tissues with pathogenic microorganisms, and in such cases the technique is not recommended to avoid contamination. Moreover, in some infectious diseases, there is an acceleration of tissue degradation, affecting the preservation of tissues and reducing the time for manipulation. One example is infection by *Clostridium chauvoei*, a bacterium when multiplied produces a toxin that causes injury to the host body as well as muscles and other tissues. The acute disease is considered highly lethal.

After death there is limited time to work before the occurrence of the degeneration of tissues (postmortem autolysis), damaging the quality of sperm. This time must be sufficient for catching the gonads in the field and transport to the laboratory. Trained technicians are required to perform rapidly obtaining of the spermatozoa, and then adding the medium to provide substrates necessary for maintenance of gametes.

The recovery after orchiectomy is usually performed in experimental works. Once is not common execute euthanasia in animals for research, the orchiectomy is an option to simulate the death for testis and epididymis. The interruption of blood supply to the testis and epididymis causes the same degenerative changes which occur after death. The orchiectomy is indicated in any situation where you do not want to eliminate the animal. In cases of unilateral testicular involvement, contralateral testicle can be tapped.

For wildlife, the difficulty of semen collection by conventional methods such as electroejaculation and artificial vagina turns recovery from epididymis interesting. Sperm can be obtained after orchiectomy, and the animal will stay alive, but will be unable to reproduce on their own. Both techniques are considered the last chance to use male gametes. The concentration of spermatozoa recovered is limited by storage capacity in the cauda epididymis of each species. The number of insemination doses is directly proportional to this concentration.

After recovery, the epididymis can be handled in two ways:

a. Room temperature (about 19°C)

b. Chilled (4–5°C)

When an animal is found dead in the field, it was exposed to climatic conditions for hours. This ambient temperature accelerates tissue degeneration with loss of sperm viability in a given time period. This condition has been represented in previous studies to be closer to reality, to verify the time available and establish a window of opportunity to work. A large variation in ambient temperatures was reported successfully, between 18 and 24°C [2, 7–9]. At a temperature of 18–20°C, it is possible to recover viable gametes, With 41.25% of progressive motility by up to 30 hours after orchiectomy [7]. Thus, after the death of bulls exposed at Ambient temperature, the ideal time until sperm recovery is up to 30 hours. With the increase of the residence time of the gametes in the cauda epididymis after death or after orchiectomy, there is greater damage to the sperm cell. In general, motility is the first parameter affected.

On the other hand, in most studies the epididymis are kept at a refrigeration temperature of 5°C before recovery, which slows down the process of cell degradation increasing the time for collecting viable gametes. For bulls, the maintenance of refrigerated epididymis enables the achievement of viable sperm for up to 72 hours after the death [10]. By comparing the refrigerating temperature (4.9–6°C) with room temperature (21.5–17.9°C) for maintenance of the epididymis before harvesting in sheep, the highest temperature affects earlier some spermatic parameters such as the acrosome integrity, motility, concentration and morphology [11]. If possible, the testicles and epididymis should be transported to the laboratory chilled to increase the time for processing.

For spermatozoa recovery, The main techniques that can be used are:

a. Retrograde flow

b. Flotation

The methods to obtain gametes depend on the animal species, size of the epididymis and the experience of the laboratorist. The retrograde flow consists of cannulation to perfuse the lumen of epididymis with diluent, promoting the backflow of the cauda epididymis content [12]. This technique is used for large animals, which have bigger testicular size and larger epididymis. It has been used successfully in horses and cattle [6, 7]. Being a detailed technique, the time for extracting spermatozoa is wide. In general, the sperm collection technique by retrograde flow results in less contaminated samples and better sperm quality than other methods, and is more suitable [13].

For small animals, the flotation method is most appropriate because of the anatomical size of the epididymis. It consists of performing longitudinal and cross cuts in the cauda region of the epididymis, the fragments are deposited into a petri dish containing medium for release of sperm and later retrieval by filtration [14]. For the smallest mammal in the world, the shrew (*Tupaia belangeri*), this is the only technique applicable because of the epididymal size in this species [15]. The flotation method requires less technician experience, because of the facility of implementation it is also often used for large animals such as sheep [16] and bulls [5, 10]. One disadvantage of this practice is that the samples are usually contaminated with blood and cellular debris, as some blood vessels are also incised during the process. The recovered sperm concentration is lower, as some are stuck between the tissue fragments so they cannot swim through the diluent in the petri dish.

The spermatozoa collected from the epididymis are free of seminal fluid which is added by the accessory glands during ejaculation, and serves as a vehicle, stimulating sperm metabolism and provides the energy necessary for the spermatozoa to pass through the uterus [17]. For it is a gamete storage location, the concentration of spermatozoa in the epididymal tail is significantly higher than the concentration of semen after ejaculation. When sperms are retrieved from the epididymis, there is a lack of ejaculatory reflex, and thereby sperms have no contact with the plasma rich in electrolytes, fructose, ascorbic acid, various enzymes and vitamins. Because of this it is recommended to add seminal plasma or other medium that has the components necessary to maintain sperm viability after recovery. It is important that whatever the diluent, it must have composition substances favorable to sperm: macromolecules such as lipoproteins and phospholipids to act as stabilizers of the plasma membrane, nutrient source for sperm metabolism (sugars) and buffer to maintain the pH. For cryopreservation, it is indispensable to use a medium containing cryoprotectants.

Besides the absence of seminal plasma, another important characteristic of epididymal sperm is the presence of large amounts of medial and distal cytoplasmic droplets in the tail of the sperm [7, 10, 17, 18]. The cytoplasmic droplet is a small spherical mass of 2–3 m in diameter found in low amounts in the ejaculated spermatozoa as it is released during ejaculation, as shown in **Figure 1**. During the transit through the epididymis, the caudal migration of the cytoplasmic droplet occurs. The presence of distal cytoplasmic droplets is not considered a severe alteration because it does not interfere with the fertilizing capacity of the sperm.

Although it represents an alteration in the normal morphology of the spermatozoa, the presence of cytoplasmic droplets in epididymal sperm is a physiological finding. Sperms with a large number of distal cytoplasmic droplets tend to lose them after 15–30 minutes of incubation in a water bath (35°C) or after the agitation [17]. Therefore, it is recommended to keep the sperm from the epididymis in a water bath at 29 to 35°C for a period of 30–60 minutes, so that the cytoplasmic droplets are released spontaneously. With the release of the drops, the sperm changes the circular movement pattern for rectilinear.

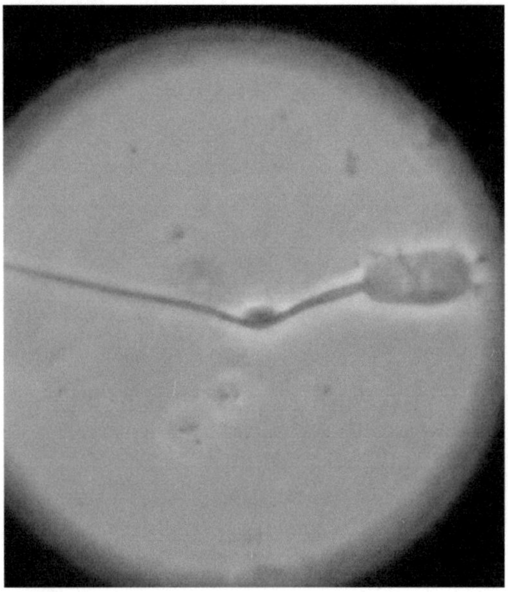

Figure 1. Phase contrast microscopy (1000×) of a medial cytoplasmic droplet in bull epididymal sperm. During the transit through the epididymis the cytoplasmic droplet migrates along the middle piece of the sperm from the proximal to the medial and distal region.

The pair of testicles (right and left sides) of a male breeder has similar characteristics of dimensions and gamete production and reserve. It has been reported for bull and stallion [6, 19]. Bulls at reproductive age (between 3 and 7 years) have epididymis (caput, corpus and cauda) with an average weight of 34.2 g. Moreover, sperm motility parameters, morphology and concentration do not differ when compared with the epididymal spermatozoa recovered from right and left. The information becomes useful in cases of unilateral testicular involvement, assisting in the reproductive male prognosis and also in clarifying the normal male genital tract physiology.

3. Cryopreservation of epididymal sperm

The first experiments arose with the need for establishment of a genetic bank for maintenance of the gametes for future use. Another reason for the cryopreservation of sperm from the epididymis is because there is a possibility that a bull could die before their semen is deposited in semen banks or at a time when only a small amount is in storage. While the use of frozen bovine semen has favored the establishment of semen banks which prolongs the usefulness of a bull and acts as a safety procedure in the event of an unexpected illness or death, not always the existing reserve is sufficient. The semen conservation methods in liquid form allow maintenance for only a few days, but it frozen in liquid nitrogen (N_2L) it is possible to keep for years.

Sperm were recovery from the epididymis of a Jersey bull aged nine years after death and were frozen for the first time on November 26, 1953 [3]. The procedure was carried out as follows: The intact scrotum from the slaughtered bull was brought to the laboratory as soon after death as possible. The time for delivery varied between 1.5 and 3 hours. On arrival the scrotal skin and tunica vaginalis were incised and the testicle was handled as aseptically as possible after its removal. Spermatozoa were recovered from the epididymis by means of several incisions in each, following which the organ was squeezed, and the expressed fluid was scraped off with a scalpel and deposited in a small volume of diluent (pasteurized homogenized whole milk and antibiotics added). Epididymal fluid containing spermatozoa were added to milk diluent. An equal volume of 20% of glycerinated milk diluent was then added gradually in increasing amounts, over a period of 30 minutes, so that a final glycerol concentration of 10% by volume was present. The sample was then placed in a refrigerator for an equilibration period of 18 hours. Subsequently the sample was distributed into ampoules and frozen.

Although some characteristics are already previously mentioned, the epididymal spermatozoa resemble the ejaculated and may be successfully cryopreserved similarly. The protocols used for freezing and thawing interfere with sperm fertility rate, so different situations are tested to identify the best protocol and the best diluent used in this process. The absence of seminal plasma in epididymal spermatozoa seems to be a positive factor in the maintenance of membrane integrity during the cryopreservation process. Furthermore, the sperm membrane composition between individual breeders can affect the resistance to cryopreservation [20]. The post-thaw incubation of the spermatozoa from epididymis of cats with seminal plasma resulted in a lower total and progressive motility, and plasma membrane integrity than control [21]. Thus, if the objective is the cryopreservation, it is recommended to add a diluent itself instead of seminal plasma.

The protocols used for cryopreservation of conventional semen with the same diluents seem to be effective for epididymal spermatozoa. What differs is the pre-preparation of sperm samples. If there is a high percentage of cytoplasmic droplets in the tail of the spermatozoa after recovery, it must remain for a period of about 30–60 minutes in a water bath, for drops release. For this first step, the sperm must be diluted in a medium without cryoprotectants that present some toxicity when stay long in contact with the sperm cells. At the end of the process, a sperm sample should be evaluated under light microscope to see if there was the release of most of the cytoplasmic droplets. After this step, the sample must be centrifuged to remove any trace of the previous medium and other dirt, and then the cryopreservation diluent with cryoprotectants is added. The amount of diluent is based on the concentration of spermatozoa and the number of insemination doses.

The epididymal spermatozoa seem to be more resistant to the cryopreservation process than the ejaculated. In sheep, the sperms from the epididymis were more resistant to the stresses caused by freezing (osmotic variations, cryoprotectants toxicity and temperature variation) compared to the ejaculate [22].

The cryopreservation protocol in liquid nitrogen and the dilution with TRIS base-egg yolk are effective [8, 23]. For bulls, the process consists of first sample centrifugation to separate the spermatozoa from other contaminants and the recovery medium. The supernatant needs to be discarded, and the pellet resuspended with an extender consisting of Tris-egg yolk and 7% glycerol as cryoprotectant. The pre-freezing parameters of motility and the total cell number should be assessed to verify the effects of centrifugation and diluent. The straws are filled with a concentration of 20 million viable spermatozoa and sealed. For temperature stabilization, the doses were maintained for 3 hours in a semen cooling container (5°C), and then placed horizontally on a 6-cm high support in an expanded polystyrene box containing liquid nitrogen for 20 minutes. Finally, straws they are immersed in liquid nitrogen at a storage temperature of −196°C. Cryopreservation can also be performed in automatic machines with controlled temperature drop.

There is not yet a specific commercial diluent for epididymal sperm in any species. Thus, recent research was conducted in order to test and identify the best diluent in this case. Some results are mentioned below.

In sheep, the results of post-thaw viability and fertility after artificial insemination with Cryopreserved successfully epididymal spermatozoa show that the diluent with 20% egg yolk and 8% glycerol and base TES-Tris-fructose (TTF) was significantly more effective in maintaining the sperm viability [24]. Egg yolk base medium with 4% glycerol was used for epididymal sperm cryopreservation of domestic and wild cats (*Panthera tigris*) successfully, providing the application of this technology in genetic resources of banks of wild species of cats [25]. Also with the aim of maintaining a genetic resource bank of wild species, Cuvier's gazelle (*Gazella cuvieri*) spermatozoa can be cryopreserved using a diluent containing 18.5% raffinose with 20% egg yolk and 6% glycerol [26]. The concentration of glycerol varies among species and must be previously tested in each case.

The concentration of sugars and the type of carbohydrate added to the medium also affect the quality of post-thaw spermatozoa from the epididymis. In tests with epididymal sperm from wild deer (*Cervus elaphus hispanicus*), the use of monosaccharides in diluents for freezing, especially fructose, improves the maintenance of post-thaw viability compared to trisaccharides [27]. Furthermore, the addition of antioxidants such as cysteine, water-soluble vitamin analog or enzymes [28, 29] leads to improvement in total motility and post-thaw integrity in sperm plasma membranes of cat epididymal sperm.

Commercial extenders for Conventional bull semen (Botu-Bov® and Bovimix®) are viable options for cryopreservation of epididymal bull sperm. Both are effective in maintaining sufficient amounts of post-thaw viable spermatozoa for use in artificial insemination. When compared, these show no difference in sperm viability (movement, morphology and integrity of plasma and acrosomal membrane) after thawing [23]. These two extenders have egg yolk in its formulation and glycerol as a cryoprotectant.

The variations in response to cooling protocols and semen freezing must be attended. Individual male effects affect directly the longevity of spermatozoa during preservation. The mammals may be regarded as "good" or "bad" freezers, according to the characteristics of the sperm plasma membrane structure which is genetically determined and the survival predisposes to thermal stress. This fact allows researchers to sort the spermatozoa as resistant or susceptible to cryopreservation. The proportion of cells, which survive the cryopreservation protocol, is determined by the sensitivity to osmotic stress during cryopreservative addition and removal, and during cooling and rewarming. While there may be species differences in overall sperm sensitivity to cryopreservation, the ejaculate is heterogeneous with a variable resistance to osmotic stress among the cells. Nevertheless, even if we optimize the process and minimize the cell death, there will still be a proportion of cells which fail to survive. We need, therefore, to concentrate on the function of the surviving population [30].

After cryopreservation identified doses containing the spermatozoa can be deposited in canisters of cryogenic cylinder, indefinitely. Inside the cylinder, the semen is kept in liquid nitrogen (N_2L), which preserves the frozen doses at a temperature of −196°C. The storage time is indefinite as long as the same is supplied periodically in order to maintain the N_2L above the minimum required level, which should never be below 15 cm height. The cylinder should be stored and handled with utmost care, avoiding possible damage.

4. Epididymal sperm fertility

It is already proven the fertility potential of spermatozoa retrieved directly from the Epididymis in some species. The epididymal sperm, as well as the ejaculated, can be used in assisted reproductive technology (ART) as conventional artificial insemination, fixed time artificial insemination (FTAI), *in vitro* fertilization of oocytes (IFV) and intracytoplasmic sperm injection (ICSI). The viability of spermatozoa after cryopreservation depends on how well the quality is preserved throughout processing, storage and thawing procedures. High-quality semen is the one that will initiate a high percentage of pregnancies when properly used. The longest time to sperm recovery after death or after orchiectomy reduces the motility and other sperm parameters before freezing. Thus, this results in worse performance for fertilization and in consequence worse pregnancy outcomes.

In vitro fertilization is the most appropriate technique to use this material because it requires a lower concentration of viable sperm for embryo production, when compared to artificial insemination. The ability to *in vitro* and *in vivo* fertilization has been proven. Sperms collected from the epididymis tail of bulls and kept at 5°C have been used for *in vitro* production of embryos [10]. Under these conditions, it was viable to produce embryos *in vitro*, with blastocyst formation up to 9 days of development using Cryopreserved sperm from epididymides refrigerated for 24, 48 and 72 hours at 5°C. When the sperm was recovery up to 30 hours at ambient temperature (18–20°C) and then cryopreserved in liquid nitrogen successfully, it was also possible to produce viable embryos (blastocysts) of eight days from *in vitro* development. The total number of blastocysts and hatching rates were lower when the recovery of spermatozoa was performed 24–30 hours after orchiectomy. When the collection was performed 6–18 hours after orchiectomy the embryo production rate was approximately 30% [31]. This means that sooner the recovery and cryopreservation are performed, the better the results. The individual bull effect on embryo production is a relevant factor that can influence the success of IVF in those cases. Great individual variation in both post-thaw sperm parameters and embryo production between bulls can be observed. It is important to know the genetic background and the fertilization potential of sperm donors to maximize the success of IVF.

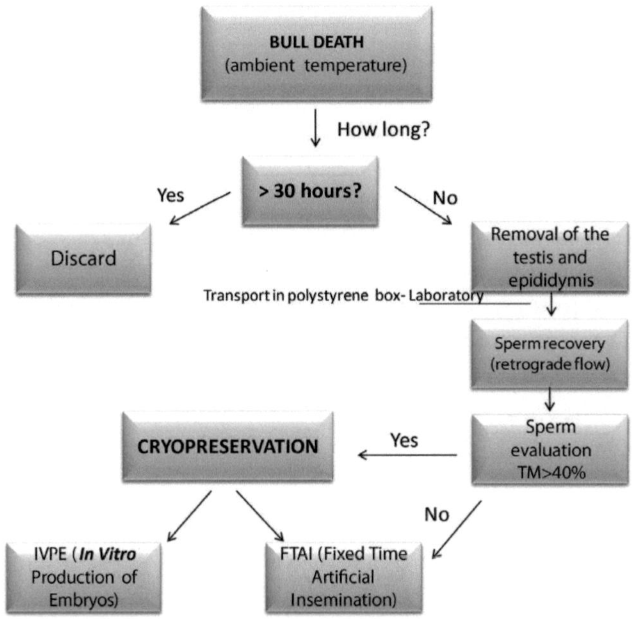

Figure 2. Flowchart of the recovery technique and sperm utilization from bulls epididymides, post-orchiectomy or postmortem when kept at ambient temperature; TM, total motility (adapted from: Bertol [34]).

Another option for use of gametes is fixed time artificial insemination (FTAI). This technique offers the following advantages: (1) does not require detection of estrus, (2) there is induction of the estrus and ovulation, (3) synchronizes births and (4) reduces the calving interval. It is already reported in the literature a pregnancy after FTAI with spermatozoa that remained for 30 hours in the epididymis at room temperature [31]. As already mentioned, recovery at room temperature is an important condition in gamete handling of epididymal reserve. After the death of a breeder, the first question we must ask is how long the animal is exposed to ambient temperature. If it was for a period of 30 hours, it would be possible to perform the retrograde flow technique for spermatozoa recovery from the cauda epididymis. If the evaluation of the total motility parameter (TM) in sperm after retrieval is greater than or equal to 40% of motile cells, it can be cryopreserved. This is important for future use without time limit in techniques of *in vitro* produce of embryos and artificial insemination. If it was less than 40% sperm must be used immediately after recovery. The flowchart of how to act in this situation is shown in **Figure 2**.

Regarding insemination with spermatozoa from the epididymis, the literature is scarce. The ability of frozen-thawed boar spermatozoa obtained from epididymides stored at 4°C for 1 day to produce piglets (three males and two females) after fallopian insemination or *in vitro* fertilization (IVF) of *in vitro* matured (IVM) oocytes was described [32]. This demonstrates the potential for *in vivo* fertilization of gametes. In sheep, insemination with the frozen/thawed epididymal sperm resulted in a lambing rate of 87.5%. Fourteen lambs (10 males, four females) were born from seven ewes. The average gestation length was 145.6 ± 1.1 days. Birth weights were 4.4 ± 0.4, 3.2 ± 0.7 and 3.6 ± 0.7 kg for single lambs, twins and triplets, respectively [18].

The technique of intracytoplasmic sperm injection to achieve fertilization, especially using retrieved epididymal spermatozoa from men with obstructive or non-obstructive azoospermia, has revolutionized the field of assisted reproduction in humans. The characteristics of postpartum development of children born after fertilization with spermatozoa from the epididymis did not differ from those from conventional methods. Children born from the intracytoplasmic injection of spermatozoa method (ICSI) with gametes from the epididymis were monitored and compared to those born from ICSI with spermatozoa from the ejaculated and by *in vitro* fertilization of embryos. The results showed that the method was successful, not to cause any fetal malformations, stillbirth or problems in the development of children [33]. These scientific findings have grounded the progress of biotechnology of reproduction. The possibility of recovery sperm from epididymis without ejaculation of animals and man, for later cryopreservation, has ensured the maintenance of important genetic source and the maintenance of compromised male in reproduction.

5. Conclusion

It is very important to have knowledge that it is possible to use the genetic reserve of a breeder even after his death. This avoids gamete wastage and an early loss of reproductive potential of a male of important genetic value. The application of this biotechnology should be proposed by the veterinarian at the time of the death of a high-value breeder. The owner of the animal should be aware that it is still possible to obtain the last descendants of the breeder. A basic protocol for the cryopreservation of epididymal sperm can be suggested as follows:

1. Recovery of gonads as soon as possible.
2. Transport to the laboratory (preferably refrigerated at 5°C).
3. Manipulation of the epididymis and the spermatozoa recovery by retrograde flow or flotation methods.
4. After recovery, spermatozoa kept in recovery medium—without cryoprotectant—in a water bath (29–35°C) for at least 30 minutes.

5. Centrifugation at least 600 × *g* for 10 minutes.

6. Resuspension with an extender of Tris-egg yolk base, and glycerol as cryoprotectant, in appropriate concentrations for each species and preparation of the straws.

7. First temperature drop to 5°C for 3 hours for stabilization, preferably in a semen cooling container.

8. Cryopreservation in liquid nitrogen by a 6-cm high support in an expanded polystyrene box or automatic cryopreservation systems.

9. Immersing straws in liquid nitrogen at a storage temperature of −196°C.

10. Storage in a cryogenic container for an indeterminate period.

It is important to note that in all of these steps (1–10), an aliquot should be removed for sperm evaluation of some parameters such as motility, concentration and morphology. This is essential for identifying flaws in the sperm handling or viability changes by temperature drop during processing, osmotic stress and the formation and dissolution of extracellular ice crystals, which can impair the fertility of sperm. The success of epididymal sperm cryopreservation depends on the effectiveness of the process, including cautious handling of the sperm cells and the technical skills in the laboratory.

Acknowledgements

The author thanks all the researchers and scientists from different institutions involved in this work, and the financial support granted by the National Counsel of Technological and Scientific Development (CnPQ) for the development of the mentioned researches from Brazil.

Author details

Melina A.F. Bertol

Address all correspondence to: melbertol@hotmail.com

Department of Technology, Postgraduate studies in Bioprocess Engineering and Biotechnology, Human and Animal Health, Federal University of Paraná, Curitiba, Brazil

References

[1] Tittarelli C, Savignone CA, Arnaudín E, Stornelli MC, Stornelli MA, de la Sota RL. Effect of storage media and storage time on survival of spermatozoa recovered from canine

and feline epididymides. Theriogenology 2006;66:1637–40. doi:10.1016/j.theriogenology.2006.01.021.

[2] Kaabi M, Paz P, Alvarez M, Anel E, Boixo JC, Rouissi H, et al. Effect of epididymis handling conditions on the quality of ram spermatozoa recovered post-mortem. Theriogenology 2003;60:1249–59. doi:10.3923/ajava.2008.400.408.

[3] Barker CAV. Low temperature preservation of bovine epididymal spermatozoa. Can J Comp Med 1954;18:390–3.

[4] Cooper TG. Sperm maturation in the epididymis: a new look at an old problem. Asian J Androl 2007;9:533–9. doi:10.1111/j.1745-7262.2007.00285.x.

[5] Martins CF, Rumpf R, Pereira DC, Dode MN. Cryopreservation of epididymal bovine spermatozoa from dead animals and its uses *in vitro* embryo production. Anim Reprod Sci 2007;101:326–31. doi:10.1016/j.anireprosci.2007.01.018.

[6] Weiss RR, Muradas PR, Graneman LC, Meira C. Freezing sperm from cauda epididymis of castrated stallions. Anim Reprod Sci 2008;107:356. doi:10.1016/j.anireprosci.2008.05.133.

[7] Bertol MAF, Romualdo Weiss R, Tomaz-Soccol V, Ernandes Kozicki L, Silva Fujita A, Azevedo de Abreu R, et al. Viability of bull spermatozoa collected from the epididymis stored at 18–20°C. Brazilian Arch Biol Technol 2013;56:777–83. doi:10.1590/S1516-89132013000500008.

[8] Turri F, Madeddu M, Gliozzi TM, Gandini G, Pizzi F. Influence of recovery methods and extenders on bull epididymal spermatozoa quality. Reprod Domest Anim 2012;47:712–7. doi:10.1111/j.1439-0531.2011.01948.x.

[9] Lone FA, Islam R, Khan MZ, Sofi KA. Effect of transportation temperature on the quality of cauda epididymal spermatozoa of ram. Anim Reprod Sci 2011;123:54–9. doi:10.1016/j.anireprosci.2010.10.012.

[10] Martins CF, Driessen K, Costa PM, Carvalho-Neto JO, de Sousa RV, Rumpf R, et al. Recovery, cryopreservation and fertilization potential of bovine spermatozoa obtained from epididymides stored at 5°C by different periods of time. Anim Reprod Sci 2009;116:50–7. doi:10.1016/j.anireprosci.2008.12.018.

[11] Lone FA, Islam R, Khan MZ, Sofi KA. Effect of transportation temperature on the quality of cauda epididymal spermatozoa of ram. Anim Reprod Sci 2011;123:54–9. doi:10.1016/j.anireprosci.2010.10.012.

[12] Garde J, Aguado M, Perez S, Garrido D, Perez-guzman M, Montoro V. Physiological characteristics of epididymal spermatozoa from postmortem rams. Theriogenology 1994;41:2003.

[13] Martinez-Pastor F, Garcia-Macias V, Alvarez M, Chamorro C, Herraez P, De Paz P, et al. Comparison of two methods for obtaining spermatozoa from the cauda epididymis

of Iberian red deer. Theriogenology 2006;65:471–85. doi:10.1016/j.theriogenology.2005.05.045.

[14] Yu I, Leibo SP. Recovery of motile, membrane-intact spermatozoa from canine epididymides stored for 8 days at 4°C. Theriogenology 2002;57:1179–90. doi:10.1016/S0093-691X(01)00711-7.

[15] Ping S, Wang F, Zhang Y, Wu C, Tang W, Luo Y, et al. Cryopreservation of epididymal sperm in tree shrews (*Tupaia belangeri*). Theriogenology 2011;76:39–46. doi:10.1016/j.theriogenology.2011.01.011.

[16] Tamayo-Canul J, Álvarez M, Mata-Campuzano M, Álvarez-RodrÍguez M, de Paz P, Anel L, et al. Effect of storage method and extender osmolality in the quality of cryopreserved epididymal ram spermatozoa. Anim Reprod Sci 2011;129:188–99. doi:10.1016/j.anireprosci.2011.11.003.

[17] Barth AD, OKO RJ. Abnormal Morphology of Bovine Spermatozoa. Ames: Iowa State University Press, 1989. 285 p.

[18] Ehling C, Rath D, Struckmann C, Frenzel A, Schindler L, Niemann H. Utilization of frozen-thawed epididymal ram semen to preserve genetic diversity in Scrapie susceptible sheep breeds. Theriogenology 2006;66:2160–4. doi:10.1016/j.theriogenology.2006.07.003.

[19] Bertol MAF, Weiss RR, Kozicki LE, de Abreu ACMR. Biometrics and sperm parameters between right and left testes and epididymis of zebu bulls. Arch Vet Sci 2015;20:156–63.

[20] Aurich JE, Kühne A, Hoppe H, Aurich C. Seminal plasma affects membrane integrity and motility of equine spermatozoa after cryopreservation. Theriogenology 1996;46:791–7. doi:10.1016/S0093-691X(96)00237-3.

[21] Thuwanut P, Chatdarong K. Incubation of post-thaw epididymal cat spermatozoa with seminal plasma. Reprod Domest Anim 2009;44:381–4. doi:10.1111/j.1439-0531.2009.01436.x.

[22] Varisli O, Uguz C, Agca C, Agca Y. Motility and acrosomal integrity comparisons between electro-ejaculated and epididymal ram sperm after exposure to a range of anisosmotic solutions, cryoprotective agents and low temperatures. Anim Reprod Sci 2009;110:256–68. doi:10.1016/j.anireprosci.2008.01.012.

[23] Bertol MAF, Weiss RR, Fujita AS, Kozicki LE, De Abreu ACMR, Pereira JFS. Two commercial extenders for cryopreservation of epididymal bull sperm. Ciência Rural 2014;44:1658–63. doi:10.1590/0103-8478cr20130747.

[24] Álvarez M, Tamayo-Canul J, Martínez-Rodríguez C, López-Urueña E, Gomes-Alves S, Anel L, et al. Specificity of the extender used for freezing ram sperm depends of the spermatozoa source (ejaculate, electroejaculate or epididymis). Anim Reprod Sci 2012;132:145–54. doi:10.1016/j.anireprosci.2012.05.006.

[25] Cocchia N, Ciani F, El-Rass R, Russo M, Borzacchiello G, Esposito V, et al. Cryopreservation of feline epididymal spermatozoa from dead and alive animals and its use in assisted reproduction. Zygote 2010;18:1–8. doi:10.1017/S0967199409990256.

[26] Martínez AF, Martínez-Pastor F, Álvarez M, Fernández-Santos MR, Esteso MC, de Paz P, et al. Sperm parameters on Iberian red deer: electroejaculation and post-mortem collection. Theriogenology 2008;70:216–26. doi:10.1016/j.theriogenology.2008.04.001.

[27] Fernández-Santos MR, Martínez-Pastor F, García-Macías V, Esteso MC, Soler AJ, de Paz P, et al. Extender osmolality and sugar supplementation exert a complex effect on the cryopreservation of Iberian red deer (*Cervus elaphus hispanicus*) epididymal spermatozoa. Theriogenology 2007;67:738–53. doi:10.1016/j.theriogenology.2006.10.005.

[28] Thuwanut P, Chatdarong K, Techakumphu M, Axnér E. The effect of antioxidants on motility, viability, acrosome integrity and DNA integrity of frozen-thawed epididymal cat spermatozoa. Theriogenology 2008;70:233–40. doi:10.1016/j.theriogenology.2008.04.005.

[29] Chatdarong K, Thuwanut P, Morrell JM. Single-layer centrifugation through colloid selects improved quality of epididymal cat sperm. Theriogenology 2010;73:1284–92. doi:10.1016/j.theriogenology.2009.12.009.

[30] Watson PF. The causes of reduced fertility with cryopreserved semen. Anim Reprod Sci 2000;60–61:481–92. doi:10.1016/S0378-4320(00)00099-3.

[31] Bertol MAF, Weiss RR, Kozicki LE, De Abreu ACMR, Pereira JFS, da Silva JJ. *In vitro* and *in vivo* fertilization potential of cryopreserved spermatozoa from bull epididymides stored for up to 30 hours at ambient temperature (18°C–20°C). Theriogenology 2016;86:1014–21. doi:10.1016/j.theriogenology.2016.03.030.

[32] Kikuchi K, Kashiwazaki N, Nagai T, Noguchi J, Shimada A, Takahashi R, et al. Reproduction in pigs using frozen-thawed spermatozoa from epididymis stored at 4°C. J Reprod Dev 1999;45:345–50. doi:10.1262/jrd.45.345.

[33] Woldringh GH, Horvers M, Janssen AJWM, Reuser JJCM, De Groot SAF, Steiner K, et al. Follow-up of children born after ICSI with epididymal spermatozoa. Hum Reprod 2011;26:1759–67. doi:10.1093/humrep/der136.

[34] Bertol, M.A.F. *In vitro* and *in vivo* fertilization potential of cryopreserved spermatozoa from bull epididymides stored for up to30 hours at ambient temperature (18°C–20°C) [thesis]. Brazil: Federal University of Paraná; 2016.

Section 4

Oocyte and Embryo Cryopreservation

Chapter 9

Cryopreservation of Embryos and Gametes: Past, Present, and Future

Julia Szeptycki and Yaakov Bentov

Additional information is available at the end of the chapter

http://dx.doi.org/10.5772/65123

Abstract

The need to find an efficient method to store gametes and embryos was driven both by medical and agricultural necessities. Gametes were the first cells used in early attempts of cryopreservation, yet these proved to be the most elusive. This chapter details the story of the development of techniques for gamete and embryo freezing, starting with hot air balloons and ending with cryotop open vitrification systems. Since gametes were the first cells to be frozen and the last to successfully thaw, their story provides an overview of the development of the science of cryopreservation.

Keywords: cryopreservation, oocytes, sperm, embryos, vitrification, slow freezing

1. Introduction

The advent of embryo cryopreservation 25 years ago [1, 2] was revolutionary as well as critical in reproductive medicine. Cryopreservation and storage of gametes and embryos provide cost and procurement efficiencies in treatment options, which otherwise would be inaccessible without substantial financial resources.

Cryopreservation maximizes fertility potential per retrieval cycle, providing a repository for individual's gametes/embryos that may not exist elsewhere. Thus reproductive potential is not limited to reproductive years, but available as one manifests the need. Use of this technology has paved the way toward single embryo transfer (SET), thereby decreasing the risk of multiple gestation pregnancy and associated health risks.

Additional application has unfolded to address needs associated with convenience and transport for the purposes of using gestational carriers, family planning, travel, and using donor oocytes/embryos in support of non-fertile couples in conception and familial continuity.

Advances in cryopreservation are paralleled and even rooted in key developments in assisted reproductive technologies (ARTs). Historically, advanced culture techniques, complex culture media, and supplements specific to the support of late stage embryonic development attempted to mimic in vivo conditions [3]. Together with carefully controlled culture environment, blastulation rates have increased exponentially from decades prior [4, 5].

In addition to these supportive measures, newer technologies including time lapse morphological assessment allow for the development of observation-based algorithms as prognostic indicators of embryonic competence. Collectively these factors form the paradigm in increased opportunities for cryopreservation and subsequent improved selection criteria for single embryo transfer. In fact, this is the current trend.

The goal of healthy ART outcome should not be clouded by commercial success rates, and while not a mandate, single embryo transfer (SET) is now widely accepted as the default position in good prognosis patients. According to a meta-analysis of randomized controlled trials, SET versus double embryo transfer (DET) in a fresh In Vitro Fertilization (IVF) treatment cycle resulted in a lower pregnancy rate, lower rate of multiple births and preterm birth, and better odds of delivering a term singleton live birth. The reported SET versus DET pregnancy rate disparity is virtually eliminated with an additional frozen SET cycle [6]. However, the immediate consequence is that in order to achieve similar results, the patient may require/need multiple cycles of embryo transfers.

Approximately 30–50% of embryos make it to a blastocyst stage. The average number of embryos frozen per IVF cycle is age dependent: women of age >35 have fewer than two embryos frozen, while younger women responding better to ovarian stimulation and producing more eggs, result in a higher likelihood of having excess embryos available for freezing [7]. The Department of Health and Human Services estimates that in 2015 more than 600K frozen embryos were stored nationwide in the USA [8]. Figures for the year 2012 released by the Human Fertilization and Embryo Authority (UK) report that of the >3.5 million embryos created since 1991, 840K (24%) were cryopreserved for clinical use. In Canada, it is estimated that >60K frozen embryos are in storage [9]. The current trend of freezing all the embryos with no fresh embryo transfer [10] in IVF treatment would suggest these numbers will likely grow much faster. Despite this uncertainty, these values underscore the importance of cryopreservation technologies.

2. Historical perspectives

Since the discovery of the tissue preserving effect of low temperatures, it has been an aspiration to maintain the vitality of human tissues by freezing. Soon after early attempts at tissue cryopreservation had failed, the main hurdle in achieving this goal became apparent; water crystallizes upon freezing and the sharp edges of the crystals disrupt cell membranes and destroy the cells. From that point onward the history of the study of cryopreservation is a description of the relentless attempts to prevent intracellular crystallization at subzero temperatures. This journey has been even more elusive for those trying to cryopreserve oocytes.

It has been known for many centuries that subzero temperatures can preserve tissue, mostly following the accidental discoveries of intact ancient animals frozen in ice for many years. However, from studying patients inflicted by frostbite it was clear that freezing may also cause tissue destruction [11].

2.1. Early days

Interestingly, the cells that were chosen for the early studies on the effects of freezing and thawing on cell viability were gametes. Spermatozoa were chosen due to their availability, small size, and their motility, which was a simple marker of viability. Oocytes were chosen since their size is large enough to allow for simple morphological evaluation.

Spallanzani, in 1776, was the first to study the effect of subzero temperatures on stallion semen and silkworm eggs [12]. He discovered that when thawed the sperm regained its motility. This was the first report of a successful sperm freeze thaw. However, it was not until 1938 that Jahnel, while searching for a remedy for syphilis, found that sperm cooled to −79°C for 40 days regained some of its motility upon thaw and reinvigorated efforts to devise an efficient freezing method [13].

Very early in the study of cryopreservation two opposing schools of thought had been developed in parallel; slow freezing with gradual desiccation of the cell and ultra-rapid freezing of small volumes also known as vitrification.

2.2. Vitrification

The term vitrification originates from the Latin word "vitreum" (glass) that describes the transformation of a substance into a non-crystalline amorphous solid. The process commonly involves rapid cooling of a liquid so that it passes through the glass transition to form a vitrified solid.

The French, Joseph Luis Gay-Lusac, in 1804 ascended in a hot air balloon and noticed that the water drops (size around 8–10 μm) in the clouds are not frozen despite the sub-zero temperatures (−5°C) [14]. He later went on to find that water can be subcooled to −12°C when contained in small tubes [15]. In 1858, Albert J.R. Mousson, a Swiss physicist, had found that the smaller the sprayed water droplets (diameter < 0.5 mm), the longer they can stay subcooled [16]. The liquid state of the water droplets in subzero temperatures is attributed to rapid cooling forming a non-crystalized solid. Luyet in his book coined the term "crystallization zone", which relates to the range of temperatures in which crystals form. He concluded that in order to avoid crystallization one must traverse this zone faster than the time it takes to form crystals [14]. It later became apparent that the small volumes are essential for achieving the high cooling velocity since it is proportional to the ratio of surface area to volume. It was also noticed that different solutions of similar volumes cool at different rates. The concentration of the solutes was shown to affect the "thermal mass (heat capacity)", which represents the ability of a substance to store thermal energy and is inversely proportional to the velocity of cooling. Pure water has a very high heat capacity and therefore is almost impossible to cool fast enough to exceed crystal growth unless very small volumes are used [17]. Walton and Judd measured the velocity of ice crystal growth in undercooled water and found it to be 65 mm/s, thereby providing the basis for calculation of the necessary speed of cooling to avoid crystallization [18]. Fahy and Rall found that in order to vitrify pure water a cooling rate of 100×10^{6}°C/min is necessary. Since such cooling velocities are not feasible, to achieve vitrification one needs to increase to solute concentration (cryoprotectants) and reduce the solutions volume. This is the current basis for clinically applied vitrification [19].

The work done by Luyet and Hodapp with colloid solutions (gelatin or agar) had led to the first successful vitrification of sperm [20]. They were able to show that the water content of the solutions was the determining factor on whether vitrification was achievable. With a 50% gelatin solution they were able to vitrify layers of 0.3 mm; however, when using a 10% gelatin solution, they could only vitrify a layer a few microns thick [15]. The drawback of these concentrated solutions was their cell toxicity. Therefore, there is a need to balance the solutions' cooling velocity on one hand and the solutes' cell toxicity on the other. It was not until 1985 that an ice-free cryoprotectant system was developed that could attain vitrification and achieve live birth for vitrified thawed mouse embryos [21–23]. Others were able to achieve high post-thaw survival rates with vitrified hamster oocytes, as well as with immature and mature murine oocytes [24–26].

Attempts to simplify the vitrification solution using a high concentration of a single cryoprotectant (dimethyl sulphoxide, DMSO) were initially successful for mouse and hamster oocytes, but later proven to be toxic causing aneuploidy, malformations and a high rate of miscarriage [27–30]. These publications halted further attempts to vitrify oocytes and focused the attention on the alternative, slow freezing.

2.3. Slow freezing

Parkes et al., in 1945 discovered, accidentally, that the rate of cooling is associated with post-thaw survival rate. They found that large containers used for freezing semen, in which, due to the large volume, the rate of cooling is slower, gave the best post-thaw motilities [31]. Hence, opposite to vitrification, slower cooling rates were associated with better cell vitality. The explanation for this observation was the physical principle of osmotic dehydration; as ice crystals formed in the suspending solution, the relative concentration of solutes in the unfrozen fraction of the solution increased and thereby increasing its osmolality. The cells suspended in the solution will respond to the higher osmolality by losing water. Therefore, slower cooling rates are associated with greater cellular dehydration and reduced risk of intracellular ice crystals formation, leading to a better post-thaw viability. Further work by Chang on rabbit ova recognized the importance of cooling rate on the maintenance of viability, the artificial activation of oocytes by rapid cooling and the achievement of litters from embryos stored at 0°C [32, 33].

Mazur was the first to describe cell-specific optimal cooling rates [34]. He was able to formulate an equation that was based on the rate at which the cells responded to osmotic pressure (hydraulic conductivity) and the effect of temperature on the movement of water across the cell membrane (temperature coefficient of water permeability) and could therefore predict cell-specific optimal cooling rates. Leibo et al. constructed a graph describing cooling rate against survival rate [35]. He showed over a 1000-fold difference in the optimal rate of cooling between oocytes (0.3°C/min) and erythrocytes (1000°C/min) due to the oocytes low hydraulic conductivity and high temperature coefficient of water permeability.

In order to guarantee ice crystal formation in the cryo-solution that will ensure the increase in its osmolality and cell desiccation, a process of ice crystal seeding was developed [36].

Two groups worked in the early 1970s independently on slow freezing of embryos. Both groups had published in 1972 the first survival of murine embryos after slow freezing [1, 2] and live offspring [1]. Both groups used slow freezing and a cryosolution containing 1 mol/l of DMSO. Wilmut and Rowson published in 1973 on the first farm animal (a calf) to be born after a transfer of a frozen thawed embryo [37].

With the advent of clinical use of IVF at the beginning of the 1980s a significant effort was made to optimize human embryo freezing in order to increase the efficiency of IVF by storing excess oocytes and embryos. This came to fruition with the first pregnancies and birth from frozen thawed embryos that were frozen using slow freezing and DMSO [38, 39]. Soon after, these were followed by publications reporting on human live births subsequent to the use of other cryoprotectants such as propanediol and sucrose. These methods proved to be more reliable and more widely adopted [40–42]. The success of human embryo freezing ignited a public debate on the ethics of embryo freezing. These ethical dilemmas prompted research on the possibility of clinical application of oocyte freezing, which was deemed to be more ethically acceptable.

In 1986, Chen reported a twin pregnancy following slow freezing of human oocytes with DMSO [43]. Chen reported high post-thaw survival, fertilization and development rates of oocytes frozen with this technique, however, attempts to replicate his success by others failed [44–46]. Furthermore, in line with the observation in animals, a high proportion of thawed human oocytes resulted in polyploid embryos [44, 47]. The poor results of oocyte cryopreservation relative to the success with embryo freezing brought clinical oocyte freezing to a halt.

2.4. Cryoprotectants

A cryoprotectant is a substance used to protect biological tissue from freezing damage. Arctic and Antarctic insects, fish and amphibians create cryoprotectants (antifreeze compounds and antifreeze proteins) in their bodies to minimize freezing damage during cold winter periods. Their exact mechanism of action is yet not fully understood. 1949, Polge et al., once again by accident, discovered the cryoprotective effects of glycerol [48]. They found that the glycerol solution protects from crystal formation during freezing by cellular dehydration. This discovery had led to successful semen storage of farm animals in 1953 and human sperm in 1964 [49]. Cryoprotectants are divided into two groups: intracellular (such as DMSO, glycerol and propylene glycol) and extracellular (such as sucrose, polyvinyl pyrrolidone, hydroxyethyl starch and dextran). One of their modes of action is lowering of the freezing point of the solution. Use of an intracellular cryoprotectant such as DMSO will prevent intracellular ice formation, while the seeding drives extracellular crystallization and the resulting increase in the osmolality of the cryosolution leading to cellular dehydration [50]. Cryoprotectants may also protect the cell membrane from the drastic changes occurring during the transition between fluid and solid states. Cryoprotectants may, however, be toxic to the cells, therefore over the years a relentless search for less toxic and efficient cryoprotectants ensued as well as for protocols combining several cryoprotectants in order to reduce individual solute concentration and the associated cell toxicity.

2.5. The return of vitrification

In 1985, Rall and Fahy were able to successfully vitrify a strew of a relatively large volume (0.25 ml) containing mouse embryos with a mixture of DMSO, acetamide and polyethylene glycol that was snap frozen in liquid nitrogen [22]. Shortly after the publication on the first births from slow-frozen oocytes, the first pregnancy and live birth from vitrified oocytes was published [51]. Developments that led to this breakthrough included the understanding that the length of exposure of the cells to the vitrification solution should be minimized to reduce toxicity [52], as well as replacing DMSO with ethylene glycol and mixtures of several cryoprotectants [53]. These changes brought about successful vitrification of bovine, murine as well as human oocytes with multiple live births [54–56]. These advancements were accompanied by the development of appropriate carriers to facilitate rapid cooling such as open-pulled straws [57], electron microscopy grids [55] and nylon loops [58]. By the end of the 1990s, vitrification was applied to human embryos achieving live births with both blastocyst and cleavage stage embryos [59, 60]. The vitrification of oocytes, despite these developments, was lagging until the introduction of appropriate carriers. The development of Cryotop in Japan was the breakthrough that allowed the adoption of oocyte vitrification into routine clinical practice. It allowed for an extremely rapid cooling rate that was facilitated by a minimal volume and resulted in a very high survival rate and live births [61–63]. A few methodological modifications that were made to the kit simplified its use and supported its wide spread distribution. Two large comparative studies established its lead role in oocyte cryopreservation [64, 65].

3. Cryopreservation protocols

Cryopreservation protocols are numerous and optimized for the cell type being frozen. These protocols fall into two major categories: equilibrium freezing and non-equilibrium freezing. Critical to either process is the partial elimination of water in the cell to avoid ice crystal damage. This chapter focuses on the two main methodologies employed in freezing reproductive cells.

Conventional slow freeze methodology is characterized as equilibrium freezing. Cells are pre-equilibrated in cryo-protecting agent (CPA) and gradual temperature depression in a controlled rate freezer optimized for the cell type being frozen is initiated. As super-cooling is achieved, a manual seeding process is required to initiate ice crystal formation outside of the cell. Continuous equilibrium of the cells is achieved by increasing the osmotic gradient initiated from the increasing proportion of ice in the surrounding medium. As a result, the cell dehydrates, thereby lowering the freezing point of the cell. At a point, with the cell being almost devoid of water, ice crystal formation is negligent and freezing occurs. This method is viewed as "forgiving" in practice, given increased pre-equilibration exposure times to relatively low concentrations of CPAs and as such promotes efficiencies by accommodating batch freezing of multiple samples.

As water excursion depends on the rate of cooling, risk can be mitigated. Rapid cooling can trap excess water inside the cell, leading to the formation of intracellular ice crystals, whereas slow cooling promotes high intracellular solute concentration by severe volume shrinkage. Both have deleterious effects on the cell.

In addition, cells that are cooled slowly are susceptible to cryo damage. Mechanisms of cryo damage include upregulation of heat and cold-shock proteins in response to cold temperatures [66, 67]. Induction of apoptosis [68], a mechanism of cryo damage, may not be immediately visible but delayed for several hours as cells try to recover from such cryopreservation stresses [69].

Largely contrasting this technology, non-equilibrium freezing was developed to overcome the many shortfalls of slow freeze methodology. Cells exposed to (usually 7.5–10%) lower strength cryoprotectant solution undergo dehydration and permeation with CPAs. Subsequent (30–60 s) rapid exposure to higher (40%) hyperosmotic solution results in complete dehydration of the cell. The sample is plunged directly into liquid nitrogen. This avoids deleterious ice crystal formation with high concentrations of CPAs and supremely rapid cooling rates (15,000–30,000°C/min). The extreme elevation in solution viscosity promotes solidification or a glass-like, suspended state as opposed to crystallization. This method requires high level manual dexterity, is labor-intensive, while offering decreased incubation times can consistently and reliably accommodate only one sample being frozen at a time. Highly skilled technicians may stagger multiple samples as per protocol, yet this leaves success rates subject to human variation. As a benefit, this method is easily introduced without the need of expensive equipment. Though unconventional, an added benefit is a comparable survival after repeat vitrification and warming of the same sample [70, 71].

Recent technological advancement into this freeze methodology is semi-automated vitrification. This platform allows simultaneous cryopreservation of up to four embryos in a closed system, addressing the long-term debate of cross-contamination in shared liquid nitrogen. Non-clinical preliminary data comparing GAVI™ (Genea BIOMEDX) to commercial manual method in mouse and donated human blastocyst stage embryos is promising [72]. Further clinical evaluation and advancement to oocytes and all embryonic stages is under way. Given success of this platform, process standardization demonstrating improved ART efficiencies may implore the few labs resistant to convert to vitrification technologies to reconsider; albeit cost considerations excluded.

4. Cryo-protecting agents

The biophysical changes that take place during temperature depression, extracellular ice formation and the creation of a potentially lethal hyperosmotic environment leads to further dehydration of the cell. Termed "freeze dehydration" this effect is implicated in organelle disruption and loss, as well as fusion or changes in cell membranes [73]. At more depressed temperatures, the viscosity of the highly concentrated solution inside and outside of the cells remains as a glassy matrix, which is relatively stable for long-term preservation.

Additional cell damage may be caused by intracellular ice formation, which is more prominent during inappropriate rapid cooling as time might be insufficient for water to move down the chemical potential gradient established by the difference in solution concentrations between the two sides of the membrane. If a cell can be cooled to a 'glass region', under conditions inhibiting ice crystal formation, successful preservation can be achieved. This is termed vitrification as previously described.

While the methodologies of slow freeze and vitrification technologies may vary within clinics, the underlying principles are fundamentally the same. Combinations of reagents provide a delicate balance between the protective and toxic effects of CPAs aiming to maintain the functional capacity of organelles, while avoiding the two main causes of cell death associated with cryopreservation: solute toxicity [74] and ice formation [34].

CPAs are generally small molecular weight solutes with high aqueous solubility, bearing polar groups that interact weakly with water [75]. CPAs act to (i) moderate the effects of the rising solute (electrolyte) concentrations in the intra and extracellular environment, (ii) stabilize intracellular protein structure and (iii) provide increasing viscosity during temperature depression that may kinetically slow or inhibit ice crystal formation. There has been a myriad of solutes that exhibit some CPA activity: amino acids (e.g. alanine, glycine, proline), amides (e.g. acetamide, formamide), diols (e.g. 1,2-propanediol, ethanediol), sugars (glucose, lactose, ribose raffinose, dextrans, hydroxyethyl starch), large polymers (polyethylene glycol, polyvinylpyrrolidone, polyvinyl alcohol) and alcohols (methanol) although some at low efficiencies [76, 77]. Modern cryopreservation protocols are largely based on few reported CPAs that are considered moderately or very effective in preserving nucleated cells [78].

CPAs can also be deleterious as osmotic effects resulting in too rapid excursion of water across a cell membrane can cause membrane rupture. Similarly, hydrogen bonding may disrupt the hydration shell around macromolecules. Chemical toxicity of high concentrations of CPAs is another cause for concern, largely in vitrification methods, the nature of which is not entirely understood [79, 80]. The suggested protein denaturation effect, even DNA conformational changes and fragmentation have been debated [80, 81]. Finally, CPAs have been shown to alter cytoskeletal components in mammalian oocytes, particularly filaments and the meiotic spindle

[82]. The reversibility of this disruption is concentration and CPA dependent and varies amongst species [54, 83–85].CPA reagents are classified as permeating, non-permeating and stabilizing.

Permeating cryoprotectants (e.g., glycerol, propane diol, dimethyl sulphoxide or ethylene glycol) cross the cell membrane through an osmotic gradient displacing water. These agents act to reduce ice crystallization and reduce cell dehydration but are toxic at higher concentrations. More specifically, in vitrification, the role is to completely inhibit ice formation. Permeating cryoprotectants also stabilize intracellular solutes which otherwise would be lethal in a hyperosmotic state. A similar dehydration effect is mimicked in the extracellular environment with temperature reduction promoting further dehydration. Dehydration is dependent upon the rate of temperature depression and limited by the cell permeability to water [35, 86].

Non-permeating cryoprotectants are generally higher molecular weight polymers (e.g., sucrose, polyethylene glycol, polyvinylpyrrolidone, ficoll, dextran). These agents mimic the dehydration mechanism of penetrating cryoprotectants but remain outside of the cell.

Generally, less toxic than penetrating cryoprotectants at the same concentration a successful vitrification strategy is to create a mixture of non-toxic level of permeating cryoprotectant(s) by the addition of non-permeating cryoprotectant. Interestingly as toxic effects of permeating cryoprotectants have been shown to be at least somewhat biochemical and unique in action, total molarity of a mixture may not be a reliable indicator of cryosolution embryotoxicity [53].

For application, CPAs are contained within a "carrier" solution that will help keep cells alive during cryopreservation. They act to provide osmotic and physiological support and avoid deviations from isotonicity, which could result in dehydration or swelling and burst of cells. It is important to note that the efficacies of carrier solutions are unpredictable and vary based on the individual or mixture of CPAs present [19].

Concern of long-term putative effects of these chemicals has paved the way for investigation into and application of extracted or modified natural biological agents, which are evolutionarily found in extreme environments. While conventional cryoprotectants interact with water, the application of uniquely acting, naturally based complementary agents, is an attractive proposition.

Biological anti-freeze molecules of sorts (e.g., cyclohexanediol and polyvinyl alcohol) [87, 88] selectively adsorb to the surface of ice crystals inhibiting ice crystal growth and ice re-crystallization. Ice blockers, including polyvinyl alcohol and polyglycerol (i.e., X-1000 and Z-1000) [89], specific to vitrification solutions act to prevent ice crystal formation. Together, these agents may also play a role in potentially damaging re-crystallization of ice growth during warming [90].

Ice-nucleating agents act to achieve deliberate ice growth in defined sites. This phenomenon is largely important in intact tissues and organs where integrated cell cooperation is essential to normal function. Unlike small tissue sections, organs are unable to effectively absorb cryoprotectant solution by simply soaking in a solution. In the case of whole organs, introduction of cryoprotectants by perfusion (through existing vasculature) is necessary. As perfusion, due to capillary distribution and time requirement of CPA diffusion, may not be equivalent throughout a larger structure, random ice crystal growth can be lethal simply by mechanical disruption. By achieving deliberate ice growth in specific sites, the damaging effects of super-cooling and likewise intracellular ice formation can be mediated and potentially avoided.

Lastly as organisms synthesize solutes and metabolites in response to cold survival strategies (e.g., trehalose [91], glycerol [92], polyols [93, 94]), understanding how biological structures interact with these mixtures may offer added benefits to current freezing regimes.

5. Slow freezing versus vitrification

To compare slow freeze technology to vitrification, the efficiencies of cryopreservation must take into consideration several factors: (i) disparities in embryo quality between the "best" freshly transferred embryo and subsequent frozen embryos; (ii) lab-specific criteria for embryo cryopreservation may foster higher implantation rates by discarding some reproductive potential of lesser quality embryos; (iii) cryo-survival should be defined in terms of complete or partial survival and (iv) post-thaw selection criteria for the transfer of cryopreserved embryos.

Therefore, a randomized control study comparing slow freeze and vitrification protocols would require standardization of protocol under optimal conditions with sibling specimens. To add more complexity in comparative analysis are individual case variations, including age discrepancies, effects of hormonal stimulation, supplementation, and endometrial priming, all of which must be taken into account.

Despite these challenges in reviewing evidence based data, as a generality the technique of vitrification has been preferentially adopted over the more traditional approach of slow cooling.

Vitrification of oocytes [95, 96] and embryos of all stages has been shown to be superior to slow freezing [6]. A large amount of clinical data suggest that one of the major consequences of the intracellular damage to embryos from slow "conventional" freezing is decreased survival as well as diminished implantation potential and outcomes when compared to vitrification [97–99]. Despite lower survival rate, there are some data that suggest similar if not improved implantation rate with slow freeze technology with fully intact good quality day 3 embryos [100]. It is known that embryo survival is not an all or none phenomenon, and therefore, comparison should be stratified on a similar quality basis.

The lack of homogeneity in some reported data is anticipated and may be due to laboratory practice or clinic-specific differences, as with other ART procedures.

As the majority of early vitrification was with cleavage stage embryos, it was recognized that failure to develop to an expanded blastocyst stage was largely a consequence of chromosomal compromise and inability to lead to a successful outcome. A bifurcated movement to karyotype embryos through pre-gestational genetic screening and cryopreserve blastocysts rather than their cleavage stage counterparts is advantageous in the identification of embryo competence and in reducing risk of miscarriage and chromosomal defects [101, 102]. This practice is important in that common morphological parameters of blastocyst scoring are not related to chromosomal status [101] and particularly for women of advancing maternal age [103].

Over the past decade, with vitrification, it has become a standard of practice to expect a post-thaw survival of >90% [104, 105] and implantation and pregnancy potentials marginally equivalent to fresh embryos [106–109].

Studies reveal longer gestational periods and heavier and healthier babies born as a result of frozen embryos compared to their fresh counterparts [110]. It is not clear whether this is related to cryotechnique or maternal factors [111]; but confirms the value of vitrification.

Reports of increased post-revitalization implantation potential over fresh counterparts [112] may be a consequence of staggered embryo transfers in which embryo procurement and implantation are performed in separate cycles. In this scenario, optimal synchronization and endometrial receptivity may be achieved in contrast to the impact of high levels of hormones present in harvesting cycles [113].

Success rates with vitrification supporting this revolutionary technology are not limited to gametes and embryos; however, extend to gonadal tissues, and non-reproductive applications including cornea [114], brain [115], heart [116], vascular [117] tissues, and cartilage [118]. The permeation of larger tissue sections and even whole organs, e.g., ovaries [119–122], shows promise in transplantation. Efficacy and potential of vitrification technologies as demonstrated through such a broad spectrum of applications justify its utility and warrant further investigation into enhanced cryopreservation potential.

Though the safety and efficacy of cryopreservation technologies is largely supported by current success rate, however, some degree of uncertainty and challenge remains.

Human embryonic stem cells (hESCs) have been established from isolated inner cell masses and more recently from single blastomeres obtained from cell stage embryos [123, 124]. The systemic reporting of chromosomal abnormalities and the recurrent manner in which they appear highlights the importance of understanding the underlying source [125]. In part, these changes are ascribed to the cryopreservation method, "adaptive pressure to" *or* "lab-specific variations in" cell culture [126–128] or are simply inherent to the cell itself [129–131]. Similarly, IVF embryos may be associated with increased risk of epigenetic abnormalities. At least in the case of hESCs, for cell line stability and quality assurance, the safety and efficacies of different cytogenetic methodologies have been assessed as they relate to genomic integrity and chromosomal stability [132]. As chromosomal instability is largely related to carcinogenesis, similar investigation into embryo cryopreservation methods may provide insight into the quality and safety of established cryopreservation protocols. Understandably, in as much as embryonic culture periods are acute in length (as compared to hESCs), still, the long-term effects of even small epigenetic changes are unknown.

6. Trends in embryo storage

Worldwide, 1.5 million ART cycles are performed each year and this number continues to rise [133]. Up to a third of patients who undergo IVF have supernumerary embryos that are cryogenically stored. A case is made for the perpetually increasing reserve of embryos and more importantly those of patients lost to follow up. For reasons undescribed here, these embryos are termed "abandoned".

At least in the USA, estimates provide that there are up to 1.4 million abandoned embryos [134] and though unknown it is safe to assume this number dwarfs the combined worldwide total. Even with permitting signed patient consents, clinics are hesitant to act and discard unclaimed embryos, largely because of the lack of regulatory guidance [135], leaving clinics vulnerable to unanticipated legal ramifications. Recently, several solutions have been proposed.

Consideration of imposing strict time limits on storage, outlining relevant responsibilities of fertility clinics and patients, and clarifying absolute guidelines related to unrestricted utility of embryos in terms of donation for third party, teaching or research use must be defined within a strict legal framework. Above all adequate long-term storage facilities are lacking.

A somewhat limited solution supporting non-fertile couples in conception and familial continuity is embryo donation. It is noteworthy to mention that cryopreservation has indirectly found a place in a larger market in terms of the transport and exchange of all types of biological samples. Initial concern over the sensitivity of microscopic volumes employed in vitrification to potentially shifting conditions during shipment has largely been overcome by advanced vapor shipping dewars, temperature monitoring, and precautionary handling.

Specific to embryo donation, ethical and genetic consequences of donation to related and unrelated parties, including offspring, must be carefully considered. In addition, given lack of restrictions on storage time limits, decades old donated embryos, thus far, may provide relatively antiquated genetics which ultimately interfere with the natural evolution of the population.

Similarly, in the case of multiple embryos donated to different parties, even a marginal risk of unsuspecting, related siblings, procreating by chance may have devastating consequences for the developing fetus. This is even more likely if embryos are donated to a clinic within the same geographical region. As such, advanced screening methods prior to procreation and/or pre-natal testing may be of benefit.

Contrary to this line of thought, in a separate context, in animal and cell line research laboratory settings, recessive gene expression and cell line mutations confer an advantage for study purposes and cryopreservation provides cell line stability against undesired changes induced by adverse events or long-term culture [136].

http://dx.doi.org/10.5772/65123

7. The thawing process

Given the worldwide exchange of reproductive cells, which is practically commonplace, IVF labs generally house a repository of embryos cryopreserved with various methodologies and formulations. The core responsibility of clinics is to ensure safety and best practice outcomes, and this requires staff training and laboratory access to a myriad of cryopreservation formulations and techniques.

In actual practice, IVF laboratories may permit revitalization of embryos using readily available thaw solutions/protocols. This is in contrast to purchasing the specific formulations matched to the cryopreservation solutions that the cells were frozen in. The impact of such mix and match freeze/thaw practices on embryos/oocytes is largely unexplored and may be quite significant given the unique actions of CPAs and unpredictability of carrier solutions as previously described. Observed success may be due to the robustness of certain reproductive cells and this may provide some artificial confidence in this regard. Prior to such validation cross-use of combinations of cryoprotectant solutions should be approached with caution.

The development of cryopreservation techniques has had immense impact across many disciplines, most notably reproductive medicine. While significant advances have been made, further advances are needed in the changing landscape of fertility. Cryopreservation has allowed IVF to evolve into a medical procedure that is efficient, safe, readily accessible, and relatively affordable.

8. Oocyte cryopreservation

While the cryopreservation of cleavage stage embryos has been a proven method for 20 years, more recently, all pre-implantation stages of embryos, including oocytes, were shown to be successfully frozen. Recent resurgence into oocyte freezing makes this application a noteworthy aspect of cryopreservation as it applies to clinical reproductive medicine.

Oocyte cryopreservation was initially focused on fertility preservation of females undergoing gonadotoxic treatments. Further application circumvented restrictions imposed on embryo freezing, which were largely a consequence of ethical, moral and legal boundaries barring embryo cryopreservation. Similarly, viewed as insurance for individuals of advanced reproductive age, oocyte banking for non-medical purposes, otherwise termed social egg freezing, supports future fertility potential in an increasingly growing group.

Women delaying childbearing until age 35 are a growing group. A trend citing increasing pregnancy rates in women in their thirties and forties is attributable to first births rather than subsequent birth, and is more pronounced in women of higher education. Psychosocial issues supporting delayed conception include parental financial stability, decreased marital discord, and increased behavioral and cognitive test scores of offspring. These are compounded with increased risk to mother and fetus including prenatal care requirements, fetal distress, preterm birth, neonatal intensive care admission, and morbidity to women of advanced maternal age [137–140].

In the USA, first birth rates for women aged 35–39 generally increased from the mid-1970s to 2012, while steady increases for women aged 40–44 began later in the early-1980s [141]. While the trend is not as dramatic in Canada, over the past two decades the average age for births to first time mothers in 2011 had risen to over 30, the oldest age on record. The year 2015 marks the first year that the average age of British women having children has passed 30 for the first time. More women over 35 are now first time mothers than the under 25s in marked contrast with the pattern as recently as 5 years ago [142]. This trend of women waiting longer to have children is consistent across race and ethnicity [143].

The high success rate with egg donation confirms that egg quality, rather than uterine factors associated with age, is the primary barrier to pregnancy in older women [144]. Progressively by early 40s to age 43, the chance of becoming pregnant through IVF exponentially decreases to near 5% and by age 45, the use of donor eggs is the only reasonable alternative. Despite these dismal outcomes, many couples or single women in their early 40s will choose to accept the lower chance of becoming pregnant and use their own eggs.

Egg freezing for preservation of fertility shows promise for success. Age remains a problem faced by women interested in using elective egg freezing. As the age of women undergoing egg freezing increases, the outcomes of assisted reproductive technology cycles utilizing their frozen eggs become less favorable.

A non-discriminatory cost-basis analysis of otherwise healthy 25-year-old women foregoing fertility until 40 revealed oocyte cryopreservation as cost-effective if IVF cycles exceeded $22,000 [145]. A hypothetical decision tree surrounding elective oocyte cryopreservation with procreation attempt at 3, 5, or 7 years after initial decision reveals greatest improvement in probability of live birth occurring if oocytes are banked at 37; noting an additional $29,000 cost per live birth in this group otherwise. However, highest probability of live birth was achieved with oocyte cryopreservation <34 years of age with no cost benefit observed for 25–30 year old age range delaying pregnancy to 40 years of age [146]. A separate analysis cites 36 years of age as the upper cut-point of non-donor oocyte cryopreservation for "success versus failure", with vitrification technology superior to slow freeze methodology [147]. Although an absolute value may not be identified for childbearing based on individual factors and resources, success probabilities at 42 years of age declining to <5% may safely advocate against oocyte cryopreservation for women >42 years of age. These models may not be reflective of all patient populations including elective, infertile, and cancer patients pursuing oocyte freezing, and individualized analyses may provide a more discriminatory framework.

Fertility preservation for (non-)medical reasons is controversial and becoming increasingly common [148]. Ethicists have upheld women's reproductive freedom while pointing out that the so-called social freezing merely postpones social problems, rather than solving them. The real challenge is two-fold. There is a clear lack of information and inadequate regulation.

http://dx.doi.org/10.5772/65123

Success rates of frozen oocytes vary among clinics, and this is reflected in conflicting statistics and the lack of a scientifically sound framework for patient education. Reports are as low as 10 and as high as 60% success rate. Access to data in establishing clinic-specific reliable predictors is lacking as individual clinics are limited in critical mass numbers to effectively determine the feasibility of this relatively new technological offering.

Though it has been reported that rates of survival, fertilization, and implantation of "young" cryopreserved oocytes fertilized with ICSI are comparable to those of fresh oocytes [149], limits imposed by nature are a constraint lending to advance planning as egg quality decline begins at age 30 and increases significantly after age 35. With respect to aging, this technique of suspending the biological clock aiming to reconcile "personal and professional timelines" must align with current limits of scientific technologies and should be critically discussed on a case-by-case basis. Critics warn of bio-objectification [150], where women could be even considered unaware victims of "a commercially exploitative context, thus undermining rather than expanding reproductive autonomy" [151].

An acceptable degree of success allowed regulatory bodies providing ART oversight including ASRM, CFAS to lift the experimental designation of oocyte cryopreservation; albeit with limited guidance. Still the majority of health care companies have yet to support elective oocyte cryopreservation for purposes other than medical necessity. Select companies are leaders in providing paid benefits for social egg freezing [152]. Mollifying procreation with career casts light on the authenticity of this offering as the employment organization benefits by not prioritizing the adjustment of the social framework of the employment organization to incorporate motherhood. Rather opinions suggest these companies seek "a productive, not a reproductive, workforce" [153].

This controversial interaction between technology and society shifts the attention from a medical procedure to a social phenomenon, which needs to be analyzed within a regulatory framework of bioethics, biopolicy, bioeconomy, and biolaw [154] with unbiased, validated reporting. In this regard, men and women can make educated choices in life decisions to harmonize personal, professional needs [155], and pregnancy.

Author details

Julia Szeptycki[1] and Yaakov Bentov[2*]

*Address all correspondence to: Yaakov.Bentov@gmail.com

1 TRIO Fertility, University of Toronto, Toronto, ON, Canada

2 ANOVA Fertility, University of Toronto, Toronto, ON, Canada

References

[1] Whittingham DG, Leibo SP, Mazur P. Survival of mouse embryos frozen to −196° and −269°C. Science 1972 Oct 27;178(4059):411–4.

[2] Wilmut I. The effect of cooling rate, warming rate, cryoprotective agent and stage of development on survival of mouse embryos during freezing and thawing. Life Sci II 1972 Nov 22;11(22):1071–9.

[3] Chronopoulou E, Harper JC. IVF culture media: past, present and future. Hum Reprod Update2014 Jan-Feb;21(1):39–55.

[4] Jones GM, Trounson AO, Gardner DK, Kausche A, Lolatgis N, Wood C. Evolution of a culture protocol for successful blastocyst development and pregnancy. Hum Reprod1998 Jan;13(1):169–77.

[5] Menezo Y, Veiga A, Benkhalifa M. Improved methods for blastocyst formation and culture. Hum Reprod1998 Dec;13 Suppl 4:256–65.

[6] Edgar DH, Gook DA. A critical appraisal of cryopreservation (slow cooling versus vitrification) of human oocytes and embryos. Hum Reprod Update2012 Sep-Oct;18(5): 536–54.

[7] Sherbahn R. Embryo freezing after IVF: Human blastocyst and embryo cryopreservation and vitrification. 2012; Available from: <http://www.advancedfertility.com/cryo.htm>.

[8] Eckman J. The Ethical Dilemmas Associated with Frozen Embryos 2015; Available from: https://graceuniversity.edu/iip/2015/05/the-ethical-dilemmas-associated-with-frozen-embryos/.

[9] CBC. Prospective Canadian Parents Turn to U.S. Surplus Embryo Donation. 2014; Available from: http://www.cbc.ca/news/canada/manitoba/prospective-canadian-parents-turn-to-u-s-surplus-embryo-donation-1.2823981.

[10] Roque M, Valle M, Guimaraes F, Sampaio M, Geber S. Freeze-all policy: fresh vs. frozen-thawed embryo transfer. Fertil Steril2015 May;103(5):1190–3.

[11] Luyet BJ. Human encounters with cold, from early primitive reactions to modern experimental modes of approach. Cryobiology1964 Sep-Oct; 51:4–10.

[12] Spallanzani L. Observations, and experiments around freezing of sperm and eggs in humans and animals. Rome: Moderna; 1776.

[13] Jahnel F. About the regaining of activity of human spermatozoa exposed to very low temperature. Klin Wochenschr1938;17:1273.

[14] Luyet BJ, Gehenio PM. Life and Death at Low Temperatures. Luyet BJ, editor: Normandy, Mo., Biodynamica; 1940.

[15] Luyet BJ. The vitrification of organic colloids and protoplasm. Biodynamics1937;1(1): 1–14.

[16] Mousson A. A few facts concerning the melting and freezing of the water. An Pfyaft 1858;105:161–74.

[17] Burton EF, Oliver WF. The crystal structure of ice at low temperatures. Proc Roy Soc (London)1935;153:166–72.

[18] Walton JH, Judd RC. The velocity of crystallization of under cooled water. J Phys Chem 1914;18:722–8.

[19] Fahy G, Rall WF. Vitrification: an overview. In: Liebermann J, Tucker MJ, editors. Vitrification in Assisted Reproduction: A User's Manual and Troubleshooting Guide. Zug, Switzerland: Informa Healthcare; 2007.

[20] Luyet BJ, Hodapp EL. Revival of frog's spermatozoa vitrified in liquid air. Proc Soc Exp Biol Med 1938;39:433.

[21] Fahy GM, Levy DI, Ali SE. Some emerging principles underlying the physical properties, biological actions, and utility of vitrification solutions. Cryobiology1987 Jun;24(3): 196–213.

[22] Rall WF, Fahy GM. Ice-free cryopreservation of mouse embryos at -196 degrees C by vitrification. Nature 1985 Feb 14–20; 313(6003):573–5.

[23] Rall WF, Wood MJ, Kirby C, Whittingham DG. Development of mouse embryos cryopreserved by vitrification. J Reprod Fertil1987 Jul;80(2):499–504.

[24] Critser JK, Arneson BW, Aaker DV, Ball GD. Cryopreservation of hamster oocytes: effects of vitrification or freezing on human sperm penetration of zona-free hamster oocytes. Fertil Steril1986 Aug;46(2):277–84.

[25] Nakagata N. High survival rate of unfertilized mouse oocytes after vitrification. J Reprod Fertil1989 Nov;87(2):479–83.

[26] Van Blerkom J. Maturation at high frequency of germinal-vesicle-stage mouse oocytes after cryopreservation: alterations in cytoplasmic, nuclear, nucleolar and chromosomal structure and organization associated with vitrification. Hum Reprod1989 Nov;4(8): 883–98.

[27] Kola I, Kirby C, Shaw J, Davey A, Trounson A. Vitrification of mouse oocytes results in aneuploid zygotes and malformed fetuses. Teratology1988 Nov;38(5): 467–74.

[28] Liu J, Van den Abbeel E, Van Steirteghem AC. Assessment of ultrarapid and slow freezing procedures for 1-cell and 4-cell mouse embryos. Hum Reprod1993 Jul;8(7): 1115–9.

[29] Van Blerkom J, Davis PW. Cytogenetic, cellular, and developmental consequences of cryopreservation of immature and mature mouse and human oocytes. Microsc Res Tech1994 Feb 1;27(2):165–93.

[30] Wood MJ, Barros C, Candy CJ, Carroll J, Melendez J, Whittingham DG. High rates of survival and fertilization of mouse and hamster oocytes after vitrification in dimethylsulphoxide. Biol Reprod1993 Sep;49(3):489–95.

[31] Parkes AS. Preservation of human spermatozoa at low temperatures. Br Med J 1945; 2: 212.

[32] Chang MC. Normal development of fertilized rabbit ova stored at low temperature for several days. Nature 1947 May 3;159(4044):602.

[33] Chang MC. Development of parthenogenetic rabbit blastocysts induced by low temperature storage of unfertilized ova. Exp Zool1954;125:127.

[34] Mazur P. Kinetics of water loss from cells at subzero temperatures and the likelihood of intracellular freezing. J Gen Physiol1963 Nov;47:347–69.

[35] Leibo SP, McGrath JJ, Cravalho EG. Microscopic observation of intracellular ice formation in unfertilized mouse ova as a function of cooling rate. Cryobiology 1978 Jun; 15(3):257–71.

[36] Critser JK, Huse-Benda AR, Aaker DV, Arneson BW, Ball GD. Cryopreservation of human spermatozoa. I. Effects of holding procedure and seeding on motility, fertilizability, and acrosome reaction. Fertil Steril1987 Apr;47(4):656–63.

[37] Wilmut I, Rowson LE. Experiments on the low-temperature preservation of cow embryos. Vet Rec1973 Jun 30;92(26):686–90.

[38] Trounson A, Mohr L. Human pregnancy following cryopreservation, thawing and transfer of an eight-cell embryo. Nature 1983 Oct 20–26;305(5936):707–9.

[39] Zeilmaker GH, Alberda AT, van Gent I, Rijkmans CM, Drogendijk AC. Two pregnancies following transfer of intact frozen-thawed embryos. Fertil Steril1984 Aug;42(2):293–6.

[40] Lassalle B, Testart J, Renard JP. Human embryo features that influence the success of cryopreservation with the use of 1,2 propanediol. Fertil Steril1985 Nov;44(5):645–51.

[41] Renard JP, Babinet C. High survival of mouse embryos after rapid freezing and thawing inside plastic straws with 1–2 propanediol as cryoprotectant. J Exp Zool1984 Jun;230(3): 443–8.

[42] Testart J, Lassalle B, Belaisch-Allart J, Hazout A, Forman R, Rainhorn JD, et al. High pregnancy rate after early human embryo freezing. Fertil Steril1986 Aug;46(2):268–72.

[43] Chen C. Pregnancy after human oocyte cryopreservation. Lancet 1986 Apr 19;1(8486): 884–6.

[44] Al-Hasani S, Diedrich K, van der Ven H, Reinecke A, Hartje M, Krebs D. Cryopreservation of human oocytes. Hum Reprod1987 Nov;2(8):695–700.

[45] Mandelbaum J, Junca AM, Plachot M, Alnot MO, Salat-Baroux J, Alvarez S, et al. Cryopreservation of human embryos and oocytes. Hum Reprod1988 Jan;3(1): 117–9.

[46] Todorow SJ, Siebzehnrubl ER, Spitzer M, Koch R, Wildt L, Lang N. Comparative results on survival of human and animal eggs using different cryoprotectants and freeze-thawing regimens. II. Human. Hum Reprod1989 Oct;4(7):812–6.

[47] Mandelbaum J, Junca AM, Tibi C, Plachot M, Alnot MO, Rim H, et al. Cryopreservation of immature and mature hamster and human oocytes. Ann N Y Acad Sci1988;541:550–61.

[48] Polge C, Smith AU, Parkes AS. Revival of spermatozoa after vitrification and dehydration at low temperatures. Nature 1949 Oct 15;164(4172):666.

[49] Sherman JK. Synopsis of the use of frozen human semen since 1964: state of the art of human semen banking. Fertil Steril1973 May;24(5):397–412.

[50] Whittingham DG. Principles of embryo preservation. In: Ashwood-Smith MJ, Farrant J, editors. Low Temperature Preservation in Medicine and Biology. Tunbridge Wells: Pitman Medical; 1980. p. 65–83.

[51] Kuleshova L, Gianaroli L, Magli C, Ferraretti A, Trounson A. Birth following vitrification of a small number of human oocytes: case report. Hum Reprod1999 Dec;14(12): 3077–9.

[52] Ishimori H, Takahashi Y, Kanagawa H. Factors affecting survival of mouse blastocysts vitrified by a mixture of ethylene glycol and dimethyl sulfoxide. Theriogenology 1992 Dec;38(6):1175–85.

[53] Ali J, Shelton JN. Design of vitrification solutions for the cryopreservation of embryos. J Reprod Fertil1993 Nov;99(2):471–7.

[54] Hotamisligil S, Toner M, Powers RD. Changes in membrane integrity, cytoskeletal structure, and developmental potential of murine oocytes after vitrification in ethylene glycol. Biol Reprod1996 Jul;55(1):161–8.

[55] Martino A, Songsasen N, Leibo SP. Development into blastocysts of bovine oocytes cryopreserved by ultra-rapid cooling. Biol Reprod1996 May;54(5):1059–69.

[56] Yoon TK, Kim TJ, Park SE, Hong SW, Ko JJ, Chung HM, et al. Live births after vitrification of oocytes in a stimulated in vitro fertilization-embryo transfer program. Fertil Steril 2003 Jun;79(6):1323–6.

[57] Vajta G, Holm P, Kuwayama M, Booth PJ, Jacobsen H, Greve T, et al. Open Pulled Straw (OPS) vitrification: a new way to reduce cryoinjuries of bovine ova and embryos. Mol Reprod Dev1998 Sep;51(1):53–8.

[58] Lane M, Schoolcraft WB, Gardner DK. Vitrification of mouse and human blastocysts using a novel cryoloop container-less technique. Fertil Steril1999 Dec;72(6):1073–8.

[59] Hsieh YY, Tsai HD, Chang CC, Lo HY, Lai AC. Ultrarapid cryopreservation of human embryos: experience with 1,582 embryos. Fertil Steril1999 Aug;72(2):253–6.

[60] Yokota Y, Sato S, Yokota M, Ishikawa Y, Makita M, Asada T, et al. Successful pregnancy following blastocyst vitrification: case report. Hum Reprod2000 Aug;15(8):1802–3.

[61] Katayama KP, Stehlik J, Kuwayama M, Kato O, Stehlik E. High survival rate of vitrified human oocytes results in clinical pregnancy. Fertil Steril2003 Jul;80(1):223–4.

[62] Kuwayama M, Vajta G, Kato O, Leibo SP. Highly efficient vitrification method for cryopreservation of human oocytes. Reprod Biomed Online2005 Sep;11(3):300–8.

[63] Kyono K, Fuchinoue K, Yagi A, Nakajo Y, Yamashita A, Kumagai S. Successful pregnancy and delivery after transfer of a single blastocyst derived from a vitrified mature human oocyte. Fertil Steril2005 Oct;84(4):1017.

[64] Antinori M, Licata E, Dani G, Cerusico F, Versaci C, Antinori S. Cryotop vitrification of human oocytes results in high survival rate and healthy deliveries. Reprod Biomed Online2007 Jan;14(1):72–9.

[65] Cobo A, Kuwayama M, Perez S, Ruiz A, Pellicer A, Remohi J. Comparison of concomitant outcome achieved with fresh and cryopreserved donor oocytes vitrified by the Cryotop method. Fertil Steril2008 Jun;89(6):1657–64.

[66] Phadtare S, Alsina J, Inouye M. Cold-shock response and cold-shock proteins. Curr Opin Microbiol1999 Apr;2(2):175–80.

[67] Rieder CL, Cole RW. Cold-shock and the mammalian cell cycle. Cell Cycle2002 May-Jun;1(3):169–75.

[68] Baust JM, Van B, Baust JG. Cell viability improves following inhibition of cryopreservation-induced apoptosis. In Vitro Cell Dev Biol Anim2000 Apr;36(4):262–70.

[69] Baust JM, Vogel MJ, Van Buskirk R, Baust JG. A molecular basis of cryopreservation failure and its modulation to improve cell survival. Cell Transplant2001;10(7):561–71.

[70] Hiraoka K, Kinutani M, Kinutani K. Successful dizygotic twin pregnancy after recryopreservation by vitrification of human expanded blastocysts developed from frozen cleaved embryos on day 6: a case report. J Reprod Med2006 Mar;51(3):213–6.

[71] Taylor TH, Patrick JL, Gitlin SA, Michael Wilson J, Crain JL, Griffin DK. Outcomes of blastocysts biopsied and vitrified once versus those cryopreserved twice for euploid blastocyst transfer. Reprod Biomed Online2014 Jul;29(1):59–64.

[72] Roy TK, Brandi S, Tappe NM, Bradley CK, Vom E, Henderson C, et al. Embryo vitrification using a novel semi-automated closed system yields in vitro outcomes equivalent to the manual Cryotop method. Hum Reprod2014 Nov;29(11):2431–8.

[73] Mazur P. Principles of cryobiology. In: Fuller BJ, Lane NJ, Benson EE, editors. Life in the Frozen State. Boca Raton: CRC Press; 2004. p. 3–65.

[74] Kleinhans FW, Mazur P. Comparison of actual vs. synthesized ternary phase diagrams for solutes of cryobiological interest. Cryobiology 2007 Apr;54(2):212–22.

[75] Fahy GM, Wowk B, Wu J, Paynter S. Improved vitrification solutions based on the predictability of vitrification solution toxicity. Cryobiology 2004 Feb;48(1):22–35.

[76] Karow AM, Jr. Cryoprotectants--a new class of drugs. J Pharm Pharmacol1969 Apr; 21(4):209–23.

[77] Karow AM, Jr. Organ Preservation for Transplantation. Karow AM, Jr., Abouna GJ, Humphries AL, editors. Boston, MA: Little Brown & Co; 1974.

[78] Ashwood-Smith MJ. Temperature and Animal Cells. Bowler K, Fuller B, editors. Cambridge: Company of Biologists; 1987.

[79] Fahy GM. The relevance of cryoprotectant "toxicity" to cryobiology. Cryobiology 1986 Feb;23(1):1–13.

[80] Fahy GM, Lilley TH, Linsdell H, Douglas MS, Meryman HT. Cryoprotectant toxicity and cryoprotectant toxicity reduction: in search of molecular mechanisms. Cryobiology 1990 Jun;27(3):247–68.

[81] Kopeika J, Thornhill A, Khalaf Y. The effect of cryopreservation on the genome of gametes and embryos: principles of cryobiology and critical appraisal of the evidence. Hum Reprod Update2014 Mar-Apr;21(2):209–27.

[82] Vincent C, Johnson MH. Cooling, cryoprotectants, and the cytoskeleton of the mammalian oocyte. Oxf Rev Reprod Biol 1992;14:73–100.

[83] Saunders KM, Parks JE. Effects of cryopreservation procedures on the cytology and fertilization rate of in vitro-matured bovine oocytes. Biol Reprod1999 Jul; 61(1):178–87.

[84] Vincent C, Garnier V, Heyman Y, Renard JP. Solvent effects on cytoskeletal organization and in-vivo survival after freezing of rabbit oocytes. J Reprod Fertil1989 Nov;87(2):809–20.

[85] Vincent C, Pickering SJ, Johnson MH, Quick SJ. Dimethylsulphoxide affects the organisation of microfilaments in the mouse oocyte. Mol Reprod Dev1990 Jul;26(3): 227–35.

[86] Leibo SP. Water permeability and its activation energy of fertilized and unfertilized mouse ova. J Membr Biol1980;53(3):179–88.

[87] Quan GB, Wu SS, Lan ZG, Yang HY, Shao QY, Hong QH. The effects of 1,4-cyclohexanediol on frozen ram spermatozoa. Cryo Letters2013 May-Jun;34(3):217–27.

[88] Wowk B, Fahy GM. Inhibition of bacterial ice nucleation by polyglycerol polymers. Cryobiology 2002 Feb;44(1):14–23.

[89] Badrzadeh H, Najmabadi S, Paymani R, Macaso T, Azadbadi Z, Ahmady A. Super cool X-1000 and Super cool Z-1000, two ice blockers, and their effect on vitrification/warming of mouse embryos. Eur J Obstet Gynecol Reprod Biol2010 Jul;151(1):70–1.

[90] O'Neil L, Paynter SJ, Fuller BJ, Shaw RW, DeVries AL. Vitrification of mature mouse oocytes in a 6 M Me2SO solution supplemented with antifreeze glycoproteins: the effect of temperature. Cryobiology 1998 Aug;37(1):59–66.

[91] Block W. Water or ice?--the challenge for invertebrate cold survival. Sci Prog2003;86(Pt 1–2):77–101.

[92] Ring RA, Tesar D. Adaptations to cold in Canadian Arctic insects. Cryobiology 1981 Apr;18(2):199–211.

[93] Hochachka P, Somero G. Biochemical Adaptation. Mechanism and Process in Physiological Evolution. Hochachka P, Somero G, editors. Oxford: Oxford University Press; 2002.

[94] Storey KB, Baust JG, Buescher P. Determination of water "bound" by soluble subcellular components during low-temperature acclimation in the gall fly larva, Eurosta solidagensis. Cryobiology1981 Jun;18(3):315–21.

[95] Cobo A, Diaz C. Clinical application of oocyte vitrification: a systematic review and meta-analysis of randomized controlled trials. Fertil Steril2011 Aug;96(2):277–85.

[96] Levi Setti PE, Porcu E, Patrizio P, Vigiliano V, de Luca R, d'Aloja P, et al. Human oocyte cryopreservation with slow freezing versus vitrification. Results from the National Italian Registry data, 2007–2011. Fertil Steril2014 Jul;102(1):90–5 e2.

[97] Kuc P, Kuczynska A, Stankiewicz B, Sieczynski P, Matysiak J, Kuczynski W. Vitrification vs. slow cooling protocol using embryos cryopreserved in the 5th or 6th day after oocyte retrieval and IVF outcomes. Folia Histochem Cytobiol2010 Jan 1;48(1):84–8.

[98] Li Z, Wang YA, Ledger W, Edgar DH, Sullivan EA. Clinical outcomes following cryopreservation of blastocysts by vitrification or slow freezing: a population-based cohort study. Hum Reprod 2014 Dec;29(12):2794–801.

[99] Wang XL, Zhang X, Qin YQ, Hao DY, Shi HR. Outcomes of day 3 embryo transfer with vitrification using Cryoleaf: a 3-year follow-up study. J Assist Reprod Genet2012 Sep; 29(9):883–9.

[100] Zhu HY, Xue YM, Yang LY, Jiang LY, Ling C, Tong XM, et al. Slow freezing should not be totally substituted by vitrification when applied to day 3 embryo cryopreservation: an analysis of 5613 frozen cycles. J Assist Reprod Genet2015 Sep;32(9):1371–7.

[101] Jing S, Luo K, He H, Lu C, Zhang S, Tan Y, et al. Obstetric and neonatal outcomes in blastocyst-stage biopsy with frozen embryo transfer and cleavage-stage biopsy with fresh embryo transfer after preimplantation genetic diagnosis/screening. Fertil Steril2016 Jul;106(1):105–12 e4.

[102] Keskintepe L, Sher G, Machnicka A, Tortoriello D, Bayrak A, Fisch J, et al. Vitrification of human embryos subjected to blastomere biopsy for pre-implantation genetic screening produces higher survival and pregnancy rates than slow freezing. J Assist Reprod Genet2009 Nov-Dec;26(11–12):629–35.

[103] Ubaldi FM, Capalbo A, Colamaria S, Ferrero S, Maggiulli R, Vajta G, et al. Reduction of multiple pregnancies in the advanced maternal age population after implementation of an elective single embryo transfer policy coupled with enhanced embryo selection: pre- and post-intervention study. Hum Reprod2015 Sep;30(9):2097–106.

[104] Lin TK, Su JT, Lee FK, Lin YR, Lo HC. Cryotop vitrification as compared to conventional slow freezing for human embryos at the cleavage stage: survival and outcomes. Taiwan J Obstet Gynecol2010 Sep;49(3):272–8.

[105] McDonald CA, Valluzo L, Chuang L, Poleshchuk F, Copperman AB, Barritt J. Nitrogen vapor shipment of vitrified oocytes: time for caution. Fertil Steril2011 Jun 30;95(8):2628–30.

[106] Edgar DH, Bourne H, Speirs AL, McBain JC. A quantitative analysis of the impact of cryopreservation on the implantation potential of human early cleavage stage embryos. Hum Reprod2000 Jan;15(1):175–9.

[107] Guerif F, Bidault R, Cadoret V, Couet ML, Lansac J, Royere D. Parameters guiding selection of best embryos for transfer after cryopreservation: a reappraisal. Hum Reprod2002 May;17(5):1321–6.

[108] Ku PY, Lee RK, Lin SY, Lin MH, Hwu YM. Comparison of the clinical outcomes between fresh blastocyst and vitrified-thawed blastocyst transfer. J Assist Reprod Genet2012 Dec;29(12):1353–6.

[109] Roy TK, Bradley CK, Bowman MC, McArthur SJ. Single-embryo transfer of vitrified-warmed blastocysts yields equivalent live-birth rates and improved neonatal outcomes compared with fresh transfers. Fertil Steril2014 May;101(5):1294–301.

[110] Maheshwari A, Pandey S, Shetty A, Hamilton M, Bhattacharya S. Obstetric and perinatal outcomes in singleton pregnancies resulting from the transfer of frozen thawed versus fresh embryos generated through in vitro fertilization treatment: a systematic review and meta-analysis. Fertil Steril 2012 Aug; 98(2):368–77 e1–9.

[111] Pinborg A, Henningsen AA, Loft A, Malchau SS, Forman J, Andersen AN. Large baby syndrome in singletons born after frozen embryo transfer (FET): is it due to maternal factors or the cryotechnique? Hum Reprod2014 Mar;29(3):618–27.

[112] Roque M, Lattes K, Serra S, Sola I, Geber S, Carreras R, et al. Fresh embryo transfer versus frozen embryo transfer in in vitro fertilization cycles: a systematic review and meta-analysis. Fertil Steril2013 Jan;99(1):156–62.

[113] Roque M. Freeze-all policy: is it time for that? J Assist Reprod Genet2015 Feb;32(2):171–6.

[114] Armitage WJ, Hall SC, Routledge C. Recovery of endothelial function after vitrification of cornea at −110°C. Invest Ophthalmol Vis Sci2002 Jul;43(7):2160–4.

[115] Pichugin Y, Fahy GM, Morin R. Cryopreservation of rat hippocampal slices by vitrification. Cryobiology 2006 Apr;52(2):228–40.

[116] Brockbank KG, Song YC. Morphological analyses of ice-free and frozen cryopreserved heart valve explants. J Heart Valve Dis2004 Mar;13(2):297–301.

[117] Song YC, Khirabadi BS, Lightfoot F, Brockbank KG, Taylor MJ. Vitreous cryopreservation maintains the function of vascular grafts. Nat Biotechnol2000 Mar;18(3):296–9.

[118] Song YC, An YH, Kang QK, Li C, Boggs JM, Chen Z, et al. Vitreous preservation of articular cartilage grafts. J Invest Surg2004 Mar-Apr;17(2):65–70.

[119] Arav A, Revel A, Nathan Y, Bor A, Gacitua H, Yavin S, et al. Oocyte recovery, embryo development and ovarian function after cryopreservation and transplantation of whole sheep ovary. Hum Reprod2005 Dec;20(12):3554–9.

[120] Bedaiwy MA, Jeremias E, Gurunluoglu R, Hussein MR, Siemianow M, Biscotti C, et al. Restoration of ovarian function after autotransplantation of intact frozen-thawed sheep ovaries with microvascular anastomosis. Fertil Steril2003 Mar;79(3):594–602.

[121] Martinez-Madrid B, Dolmans MM, Van Langendonckt A, Defrere S, Donnez J. Freeze-thawing intact human ovary with its vascular pedicle with a passive cooling device. Fertil Steril2004 Nov;82(5):1390–4.

[122] Wang X, Chen H, Yin H, Kim SS, Lin Tan S, Gosden RG. Fertility after intact ovary transplantation. Nature 2002 Jan 24;415(6870):385.

[123] Chung Y, Klimanskaya I, Becker S, Marh J, Lu SJ, Johnson J, et al. Embryonic and extraembryonic stem cell lines derived from single mouse blastomeres. Nature 2006 Jan 12;439(7073):216–9.

[124] Klimanskaya I, Chung Y, Becker S, Lu SJ, Lanza R. Human embryonic stem cell lines derived from single blastomeres. Nature 2006 Nov 23;444(7118):481–5.

[125] Gaztelumendi N, Nogues C. Chromosome instability in mouse embryonic stem cells. Sci Rep2014;4:5324.

[126] Allegrucci C, Young LE. Differences between human embryonic stem cell lines. Hum Reprod Update2007 Mar-Apr;13(2):103–20.

[127] Hanson C, Caisander G. Human embryonic stem cells and chromosome stability. APMIS 2005 Nov-Dec;113(11–12):751–5.

[128] Imreh MP, Gertow K, Cedervall J, Unger C, Holmberg K, Szoke K, et al. In vitro culture conditions favoring selection of chromosomal abnormalities in human ES cells. J Cell Biochem2006 Oct 1;99(2):508–16.

[129] Catalina P, Montes R, Ligero G, Sanchez L, de la Cueva T, Bueno C, et al. Human ESCs predisposition to karyotypic instability: Is a matter of culture adaptation or differential vulnerability among hESC lines due to inherent properties? Mol Cancer2008;7:76.

[130] Mantel C, Guo Y, Lee MR, Kim MK, Han MK, Shibayama H, et al. Checkpoint-apoptosis uncoupling in human and mouse embryonic stem cells: a source of karyotpic instability. Blood 2007 May 15;109(10):4518–27.

[131] Peterson SE, Westra JW, Rehen SK, Young H, Bushman DM, Paczkowski CM, et al. Normal human pluripotent stem cell lines exhibit pervasive mosaic aneuploidy. PLoS One2011;6(8):e23018.

[132] Catalina P, Cobo F, Cortes JL, Nieto AI, Cabrera C, Montes R, et al. Conventional and molecular cytogenetic diagnostic methods in stem cell research: a concise review. Cell Biol Int2007 Sep;31(9):861–9.

[133] ESHRE. ART fact sheet [database on the Internet]. 2014. Available from: https://www.eshre.eu/Guidelines-and-Legal/ART-fact-sheet.aspx.

[134] Kirkey S. Put limit on how long Canadian fertility clinics can store frozen embryos, academics argue. National Post. 2016.

[135] Cattapan A. Frozen in perpetuity: 'abandoned embryos' in Canada. Reproductive Biomedicine & Society Online 2015;1(2):104–12.

[136] Bailey DW. Genetic drift: the problem and its possible solution by frozen-embryo storage. Ciba Found Symp1977 Jan 18–20(52):291–303.

[137] Jahromi BN, Husseini Z. Pregnancy outcome at maternal age 40 and older. Taiwan J Obstet Gynecol2008 Sep;47(3):318–21.

[138] Jolly M, Sebire N, Harris J, Robinson S, Regan L. The risks associated with pregnancy in women aged 35 years or older. Hum Reprod2000 Nov;15(11):2433–7.

[139] Liu K, Case A. Advanced reproductive age and fertility. J Obstet Gynaecol Can2011 Nov;33(11):1165–75.

[140] Schoen C, Rosen T. Maternal and perinatal risks for women over 44—a review. Maturitas 2009 Oct 20;64(2):109–13.

[141] Matthews TJ, Hamilton BE. First births to older women continue to rise. NCHS Data Brief 2014 May(152):1–8.

[142] Bingham J. Middle age mothers overtake the young in Britain's maternity wards. The Telegraph. 2015.

[143] Bichell RE. Average Age Of First-Time Moms Keeps Climbing In The U.S. NPR. 2016.

[144] Abdalla HI, Burton G, Kirkland A, Johnson MR, Leonard T, Brooks AA, et al. Age, pregnancy and miscarriage: uterine versus ovarian factors. Hum Reprod1993 Sep;8(9): 1512–7.

[145] Hirshfeld-Cytron J, Grobman WA, Milad MP. Fertility preservation for social indications: a cost-based decision analysis. Fertil Steril2012 Mar;97(3):665–70.

[146] Devine K, Mumford SL, Goldman KN, Hodes-Wertz B, Druckenmiller S, Propst AM, et al. Baby budgeting: oocyte cryopreservation in women delaying reproduction can reduce cost per live birth. Fertil Steril 2015 Jun;103(6):1446–53 e1–2.

[147] Cil AP, Bang H, Oktay K. Age-specific probability of live birth with oocyte cryopreservation: an individual patient data meta-analysis. Fertil Steril2013 Aug;100(2):492–9 e3.

[148] von Wolff M, Germeyer A, Nawroth F. Fertility preservation for non-medical reasons: controversial, but increasingly common. Dtsch Arztebl Int2015 Jan 16;112(3):27–32.

[149] Lockwood GM. Social egg freezing: the prospect of reproductive 'immortality' or a dangerous delusion? Reprod Biomed Online2011 Sep;23(3):334–40.

[150] Holmberg T, Schwennesen N, Webster A. Bio-objects and the bio-objectification process. Croat Med J 2011 Dec 15;52(6):740–2.

[151] Harwood K. Egg freezing: a breakthrough for reproductive autonomy? Bioethics 2009 Jan;23(1):39–46.

[152] Bennett J. Company-Paid Egg Freezing Will Be the Great Equalizer. Time. 2014.

[153] Baylis F. Left out in the cold: arguments against non-medical oocyte cryopreservation. J Obstet Gynaecol Can2015 Jan;37(1):64–7.

[154] Rose N. The Politics of Life Itself. Biomedicine, Power, and Subjectivity in the Twenty-first Century. Princeton, NJ/Oxford: Princeton University Press; 2007.

[155] Martinelli L, Busatta L, Galvagni L, Piciocchi C. Social egg freezing: a reproductive chance or smoke and mirrors? Croat Med J 2015 Aug;56(4):387–91.

Chapter 10

Cryopreservation of Human Gametes and Embryos: Current State and Future Perspectives

Jeseta Michal, Zakova Jana, Ventruba Pavel,
Bartosz Kempisty and Crha Igor

Additional information is available at the end of the chapter

http://dx.doi.org/10.5772/64950

Abstract

Cryopreservation of human gametes and embryos is an important and widely used method in most embryology laboratories. During last years, the practice of single embryo transfer was a greater demand for reliable cryostorage of surplus embryos. Currently, there are two basic principally different methods usable for cryopreservation: slow freezing and vitrification. Vitrification is a very promising method with massive use in embryology. Nowadays, this method is also suitable for cryopreservation of human mature oocytes (one of the most problematic cell in cryobiology). This progress in the field of cryopreservation opens new perspectives in assisted reproduction. Recent effective oocyte vitrification systems have a significant impact on clinical practice. This chapter gives a view of human gametes (sperms, oocytes) and embryos cryopreservation application and possibilities. Indications and methods of cryopreservation and thawing are mentioned.

Keywords: gamete, oocyte, spermatozoa, embryo, cryopreservation, assisted reproduction, vitrification

1. Introduction

Cryopreservation of human gametes and embryos is very important method in most embryology laboratories. Two basic cryopreservation techniques rule the field, slow-rate freezing (first developed) and vitrification, which have gained a foothold in recent years. Vitrification is relatively simple, requires no expensive programmable freezing equipment and uses a small amount of liquid nitrogen for freezing. Vitrification of human oocytes and embryos (especially at early stages) is more effective than slow freezing.

During last years, the practice of single embryo transfer was a greater demand for reliable cryostorage of surplus embryos. The first reports of successful freezing and thawing of human embryos were in 1983 [1]. There are a growing number of indications for oocyte cryopreservation as oocyte donation, fertility preservation for cancer patients or social egg freezing. Reproductive behaviour of women has been changed in last years. There is a delay in the age of motherhood due to various reasons like career, live style or education. It is known, that in women older than 35 years, reduction of ovarian reserve is observed. The use of younger cryopreserved oocytes can reduce the risk of foetal loss and aneuploidies associated with ageing oocytes. Oocyte cryopreservation simplifies the logistics of assisted reproductive technology (ART) cycles in donation programme, and there is no need for menstrual cycle synchronization between donor and recipient.

Damage of reproductive function is very frequent and well documented side effect associated with the treatment of malignant tumours. The increasing success of cancer treatment and determined efforts to improve the quality of life after successful treatment has turned attention to the preservation of reproductive function in young women and also in young men. Sperm freezing is largely recommended to preserve fertility prior to the oncology treatment. Cryopreservation of spermatozoa is routinely used in a variety of reasons (sperm bank, donor programme, etc.).

For this reasons, cryopreservation of gametes and embryos is more and more important part of human-assisted reproduction.

2. Cryopreservation: principles and methods

2.1. Cellular cryotolerance

Cryopreservation of cells and their storage in liquid nitrogen at −196°C is not physiological process. The freezing process can cause stress and mechanical damage of cells by ice crystal formation [2]. Cell damage can occur at any time during cryopreservation process. Cell lysis can be induced by intracellular ice formation. This major change is easily observed through routine microscopic observations. However, damages can also occur in the cellular structural/functional levels involving intracellular organelle changes, what is more difficult to diagnose.

Three types of damage during cooling process in oocytes and embryos were described:

- the damage of microtubules, including meiotic spindle with relative high temperature (from +5 to −5°C);
- intracellular crystal formation (temperature from −5 to −80°C);
- damage of zona pellucida or the cytoplasm (temperature from −80 to −150°C).

The using of cryoprotectants makes a damage of oocytes via cytotoxic and osmotic effect. The addition and the removal of cryoprotectants from the oocyte create an osmotic imbalance across the oocyte membrane, which may result in large volumetric changes and cause damages in the cell morphology, cytoskeletal structures and physiologic function.

Protein structure and function, as well as metabolism, can also be affected. Cells require a period of recovery after thawing, and then, they are able to continue normal intracellular function.

Cryopreservation affects various organelles like intracellular lipids [3], mitochondria (**Figure 1**) cortical granules [4], cytoskeletal structure, zona pellucida and also meiotic spindle [5]. Deleterious effects of meiotic spindle can resulted into chromosome disaggregation. Improper chromosome segregation could lead to aneuploidy and genetic errors, which may cause embryonic and foetal abnormalities. Furthermore, cryopreservation can induce releasing of cortical granule what makes changes of zona pellucida (zona hardening) [4].

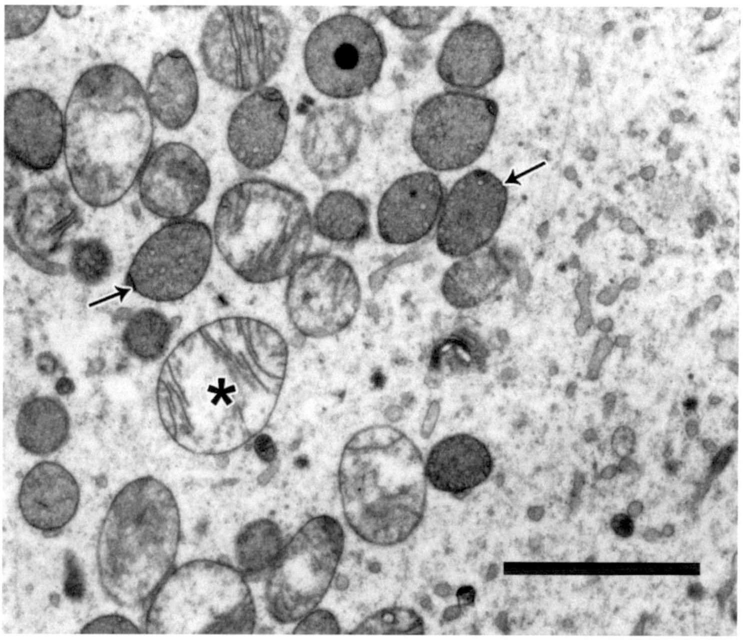

Figure 1. Mitochondria from primary oocyte after cryopreservation with morphological alterations: swollen mitochondria (*) and mitochondria with atypical tubular cristae (black arrow). Scale bar represents 2 μm.

2.2. Cryoprotectants

Cryoprotectants are substances with high solubility and cytotoxicity, what is directly proportional to their concentration and temperature. They aim to protect cells from any damage what is known as cold shock, during freezing-warming procedure. Cryoprotectants bond water and they reduce the toxic effect of high concentrations of other compounds. At high concentrations, cryoprotectants minimize the damage caused by ice formation, as they cause the water to form a glass rather than ice crystal. After thawing, the cryoprotectants must be removed from cells to avoid their deleterious effect on further fertilization and embryonic development.

2.2.1. Membrane permeable cryoprotectants

These solutions displace water via an osmotic gradient and partly occupy the place of the intracellular water. Indeed, increase in the extracellular osmolarity generates an osmotic gradient across the cell membrane-supporting dehydration of the cell. In this group, compounds are with relatively low molecular weight (<100 g/mol). The most commonly used cryoprotectants for oocyte and embryos are ethylene glycol, 1,2-propandiol and dimethyl sulfoxide (DMSO) [6]. Ethylene glycol is widely used during the vitrification of human oocytes and embryos due to its low toxicity and high permeability. In present time, it is standard part of all vitrification protocols. During equilibration step especially in oocyte vitrification, the compounds of very high concentration (>4 M concentration) are used.

2.2.2. Membrane nonpermeable cryoprotectants

Nonpermeable cryoprotectants are usually large molecules, which remain in extracellular solution. Extracellular saccharides and macromolecules (sucrose, trehalose, Ficoll, PVP) are commonly added to vitrification solutions. They help draw water out of the blastocoel to attain better dehydration and reduce osmotic shock. Very frequent approach is combination of more cryoprotectants for decreasing the individual specific toxicity of each solution. At least, one of these cryoprotectants should be permeable (with higher toxicity) and one or two nonpermeable (lower toxicity) [7].

For example, during vitrification commonly used ethylene glycol or DMSO or propanediol (permeable) are often combined with sucrose or PVP (nonpermeable), which reduce the concentration of permeable cryoprotectants and facilitate dehydration and vitrification.

Cell permeability is an important factor for determining the conditions for cryopreservation. The permeability of mouse embryos increases as development proceeds to the compacted morula. Ethylene glycol is less permeating than propylene glycol at the one cell stage. In morula stage, ethylene glycol is far more permeating than other cryoprotectants. Exchange of water and cryoprotectants in expanded pig blastocyst occurs predominantly by facilitated diffusion but in oocytes predominantly by simple diffusion [8]. This was related to the expression of aquaporin three mRNA, which was abundantly active in expanded blastocyst, but not in oocytes. The common consensus is that rapidly permeating agents are favoured for oocyte cryopreservation, because the exposure time before cooling can be shortened, and because osmotic swelling during removal of the cryoprotectant can be minimized.

2.3. Slow freezing

This technique involves stepwise programmed decrease in temperature. The procedure is lengthy and requires the using of expensive equipment (**Figure 2**). This process does not exclude ice crystal formation, which can have extremely deleterious effects [9].

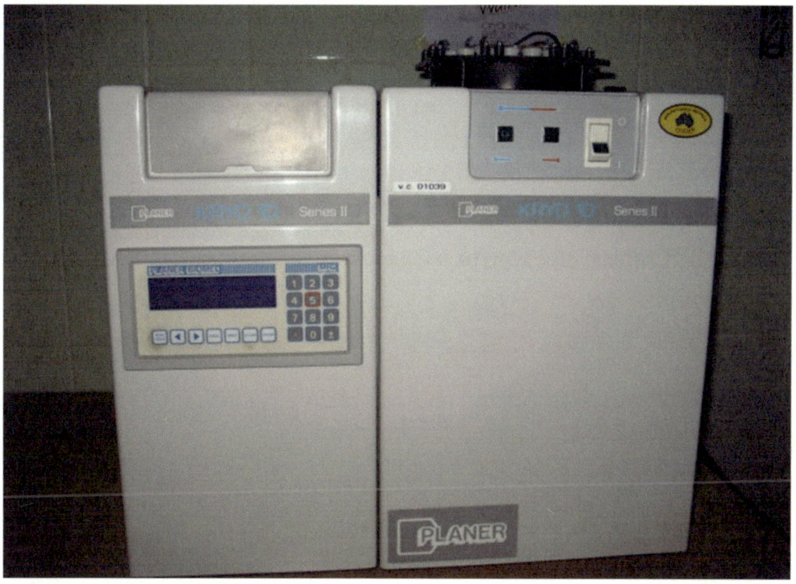

Figure 2. Programmable cryo freezer Planer Kryo F10.

Slow freezing is a technique with long history but in comparison to vitrification actually does not bring any advantages. Vitrification methods are more efficient and reliable than any version of slow freezing [10]. After a long period of practising was the convention slow freezing method completely stopped in many centres and was replaced by routine vitrification.

2.4. Vitrification

Limiting factor for all cryopreservation methods is ice crystal formation that drastically reduced survival of embryos and oocytes. Vitrification process produces a glasslike solidification of living cells, which completely avoids ice crystal formation. It is well known that vitrification requires a greater amount of cryoprotectants, what increases the toxicity of their environment. However, it was claimed higher survival rate after using vitrification instead of slow freezing [11]. Vitrification is very simple, cost-effective process, but the skills to perform require good manual training.

3. Gamete cryopreservation

3.1. Spermatozoa cryopreservation

Cryopreservation of human semen is well-established laboratory procedure to maintain the fertilizing potential of spermatozoa during storage in liquid nitrogen. Modern trends in assisted reproduction technologies influenced the indications for human sperm cryopreservation. Spermatozoa are not so sensitive to cryopreservation damage (in comparison with other cell), because of the high fluidity of the membrane and the low water content (about 50%). The effect of cryopreservation on sperm DNA integrity is still unclear. There is no agreement in literature on whether or not affect cryopreservation sperm chromatin integrity.

When clinicians became aware that azoospermia or very severe oligozoospermia could not be improved by medical treatment, it arises the idea to create sperm banks. Today, cryopreservation of spermatozoa is routinely used in a variety of reasons:

1. Donor or husband semen storage for assisted reproduction.

2. Sperm banking for husband sperm for psychological or other reasons (it is not always possible to produce sperm samples at the appropriate time in the cycle).

3. Storage of epididymal or testicular spermatozoa after MESA/TESE, to avoid repeated biopsies or aspirations.

4. Storage of sperm as a fertility "insurance" for future.

5. Preservation of semen before surgical, chemical or radiological cancer therapy, which may lead to testicular failure or ejaculatory dysfunction. Also other nonmalignant diseases, such as diabetes or autoimmune disorders, may lead to testicular damage.

6. Male gamete freezing is largely recommended to preserve fertility in those subjects who are exposed to potentially toxic agents, which may interfere with gametogenesis.

Semen preservation before the beginning of therapy should be proposed to all adult men and postpubertal boys. To date, no clinically proven methods are available to preserve fertility in prepubertal males. The testicular cancer survivors have a good chance of fathering a child by using sperm cryopreserved prior to the oncology treatment thanks to assisted reproduction methods [12].

In the ICSI era, almost all cryopreserved semen sample, even when it contains only few sperm, could be used for subsequent infertility treatment. Genetic damage is unknown.

Cryopreservation is known to cause some changes in sperm morphology, including damage to mitochondria, the acrosome and the sperm tail. The sperm motility is particularly sensitive, and it is generally accepted that it can be reduced to 50% after the cryopreservation/thawing procedure. Due to this fact, it is necessary to choose potential donors with an emphasis on this sperm parameter.

3.2. Oocyte cryopreservation

Cryopreservation of human oocyte can be an alternative to circumvent many of the ethical issues associated with embryo cryopreservation. For oocyte cryopreservation, it is very suitable to use vitrification method. Oocyte cryobanking is a new more efficient approach in oocyte donor-recipient treatment. On the basis of guideline from the Practice Committees of the American Society for Assisted Reproductive Medicine (from 2013) and, in March 2012, European Society of Human Reproduction and Embryology (ESHRE), it is indicated that mature oocyte vitrification and warming are not experimental and should no longer be considered as experimental procedures. This progress in the field of cryopreservation opens new perspectives in assisted reproduction. Recent effective oocyte vitrification systems have a significant impact on clinical practice. It is a possible way in countries where the law forbids the cryopreservation of embryos. Indeed, efficient oocyte vitrification technology eliminates synchronization between donor and recipient. It enables the establishment of egg banks by eliminating the logistics of coordinating egg donors with their recipients. Progress in oocyte vitrification brings new possibilities mainly for women, who are trying to postpone child-bearing from professional or social reasons. The process was originally developed as a way to preserve the fertility of cancer patients undergoing possibly sterilizing chemotherapy, and it is relatively simple.

Oocyte cryopreservation is less successful than embryo cryopreservation for many reasons. Oocytes have small surface to volume ratio, temperature-sensitive metaphase spindle [13], zona pellucida as very specific structure and susceptibility to parthenogenetic activation. Oocytes are one of the biggest cells with high likelihood of intracellular ice formation [14] Oocytes are very unique cells, because of their developmental capacity to be fertilized and then to support early embryonic development. This capacity derives from maternal legacy of the myriad of transcript, proteins and energetic substrates and also cytoplasmic organelles, which facilitate early mitotic divisions of the embryo until embryonic genome activation occurs [15]. This highly organized structure often incurs serious damage after cryopreservation. The volume of mammalian oocyte is much bigger than that of spermatozoa, thereby substantially decreasing the surface to volume ratio and making them sensitive to chilling and highly susceptible to intracellular ice formation. In fact, in a developing embryo, cleavage division occurs without any increase in volume until blastocyst stage, leading to higher nucleus-cytoplasmic ratio of embryo blastomeres compared with the oocyte. Oocytes are substantially more prone to cryo damage than are embryos. Number of blastomeres in early embryos provides great flexibility to compensate for any detrimental effects of cryopreservation, because missed blastomeres can be replaced by the daughter cells of dividing intact ones. Oocytes contain one-half of the genetic material of the future individual, and so any damage to its chromatin structure may result in deleterious defects in the developmental competence of the resulting embryos. Damage of meiotic spindle can result in chromosomal abnormalities after thawing. The permeability of oocyte plasma membrane to cryoprotective agents is low compared with embryo [6].

Although mature oocytes in metaphase II are sensitive to cryopreservation (detrimental effect on meiotic spindle or premature cortical granule release) and immature oocytes on prophase I (GV oocytes) look that are more suitable for cryopreservation. It is well known that oocytes frozen at GV stage exhibited decreased affectivity of *in vitro* maturation and increased spontaneous parthenogenetic activation [16]. For this reason in case of immature oocytes, it is recommended to use *in vitro* maturation and after that perform their subsequent vitrification.

It was presented that highly organized structure of fresh oocyte changes dramatically (at cellular, ultrastructural, molecular and developmental levels) after cryopreservation. Cryo-preserved oocytes have cellular characteristics that differ from those of the fresh oocytes.

3.2.1. *Cryopreservation of ovarian tissue*

Fertility preservation has a great importance to many young women with cancer [17]. Cryopreservation of ovarian tissue is a safe, simple and effective option for preserving fertility in young patients facing or undergoing gonadotoxic therapy. Oocytes in primordial follicles are very small and tolerate cryopreservation very well. The removal of ovarian tissue is a simple procedure. Ovarian tissue can be obtained using minimally invasive techniques during laparoscopy, with unilateral ovariectomy or partial ovariectomy. Ovarian tissue can be cryopreserved independently of the menstrual phase.

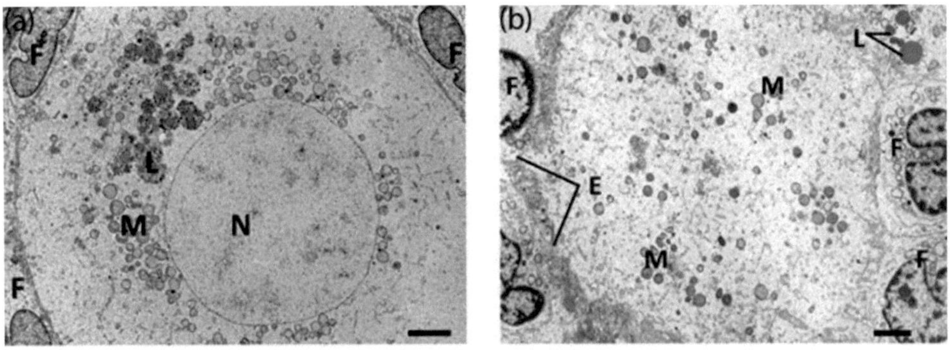

Figure 3. Primary follicle from ovarian cortex before (a) and after (b) cryopreservation with morphological alterations. The oolemma of oocyte after cryopreservation is more undulated and interrupted (E), and the cytoplasm of follicular cells (F) is vacuolated. N, nucleus; M, mitochondria; L, lipid droplets. Scale bar represents 5 μm [19].

In 2004, first live birth after autotransplantation of human ovarian tissue was reported [18]. To date, 60 live births have been reported worldwide following transplantation of cryopreserved ovarian tissue. However, research on the cryopreservation of ovarian tissue as a method of fertility preservation has now been continuing for more than a decade, and considerable successes have recently been achieved.

In centres that offer cryopreservation of ovarian tissue, the procedure can be performed one day after the patient's first visit. After the tissue has been removed, it can be processed immediately or transferred in special transportation containers to a centre specializing in the cryopreservation of ovarian tissue, with an associated cryobank (**Figure 3**).

4. Zygote and embryo cryopreservation

Cryopreservation of human embryos is a safe procedure, which has been carried out for more than last 30 years. In development of *in vitro* techniques and together with single embryo transfer becoming greater demand for an efficient and reliable cryopreservation method for surplus embryos. It is possible to cryopreserve the human zygotes immediately after fertilization, at the pronuclear stage or embryos during early cleavage stages (2–8 cells) or at the expanded blastocyst stage (after 5–7 days in culture).

Embryos are cryopreserved in any embryonic stages. Still there does not exist a common consensus what is the most optimal developmental stage for embryo cryopreservation.

Since morphology of vitrified and thawed embryos is not enough to assess the viability, the possibility of culturing for a few more days before transfer can ensure that embryo is for transfer. In contrary to oocytes, embryos are after cortical reaction, which gives the ooplasmic membrane more stability to cope with the low temperature and osmotic changes.

4.1. Cryopreservation of zygotes

For cryopreservation of human zygotes, it is suitable to use only vitrification method. Slow freezing method has more than threefold worse results than vitrification [20].

The recent reported data for successful pregnancies suggested that the vitrification of human zygotes and early-stage embryos is a perfect alternative to slow freezing techniques especially in countries where cryopreservation of later stage human embryos is prohibited either by law or due to religious reasons.

4.2. Cryopreservation of cleaved embryos

Vitrification of early-stage human embryo is acceptable and better alternative than slow rate freezing because of the higher survival rate and increased rates of pregnancy. Cryopreservation of cleaved embryos is not so effective as cryopreservation of blastocyst.

4.3. Cryopreservation of blastocyst

Blastocysts are the top embryos, what have successfully passed the critical step of genomic activation and have a high developmental potential. Their advantage is containing numerous small cells; thus, the loss of some cells during freezing and thawing is probably less harmful for future development of the embryo. Furthermore, during extended cultivation, embryos with worse viability are arrested in development and will not be cryopreserved.

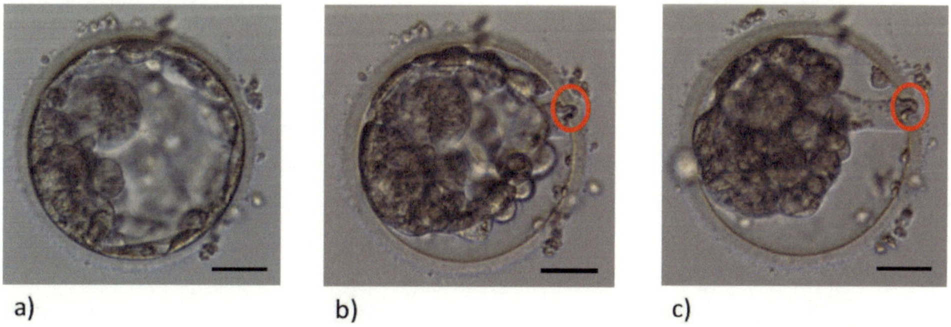

Figure 4. Laser blastocoel puncture (assisted shrinkage): human expanded blastocyst before puncture; (a) laser pulse open zona pellucida (red circle) and make a small defect in the trophectoderm (b), which resulted to blastocyst shrinking (c). Scale bars represent 30 μm.

Blastocyst presents special challenge to cryopreservation. Excessive water in the blastocoel may lead to ice formation and subsequent damaging of cellular structures. To minimize this risk, removal of some of the blastocoel fluid has been attempted. Removal of blastocoel fluids can be done by perforating the blastocoel and letting the fluid flow passively out [21]. The process called assisted shrinkage can be performed in a variety of ways, including microneedle puncture, repeated micropipetting of the blastocoel or laser-pulse opening of zona pellucida (**Figure 4**).

5. Our experiences with cryopreservation of gametes, embryos and tissues

We have cryopreserved sperm since 1991, and we introduced embryo freezing in 1995. In the beginning, we performed slow freezing by Planer, but from 2007, we prefer vitrification. Well-functioning cryopreservation was an essential prerequisite for a donation of gametes and embryos program. Within the centres of assisted reproduction, we had the first and the largest sperm bank in the Czech Republic from 1995 (currently, we have 100 donors).

We have also built a centre for fertility preservation for both male and female oncologic patients. Methods of preserving the fertility in young women can be divided into three cryopreservative methods: embryo-, mature oocyte- and ovarian tissue-cryopreservations. We have started as the first with ovarian tissue freezing and sperm freezing before gonadotoxic treatment in the Czech Republic. We cryopreserved ovarian tissue of 23 women before gonadotoxic treatment (from January 2006 to December 2015). During October 1995 to December 2015, we cryopreserved the sperm of 1231 men—oncologic patients (587—testicular cancer diagnosis). The testicular cancer survivors have a good chance of fathering a child by using sperm cryopreserved prior to the oncology treatment, even when it contains only limited number of spermatozoa. There are 41 patients in our centre, who returned for infertility treatment underwent 58 treatment cycles with cryopreserved sperm. Totally, 20 pregnancies were achieved, that is 34.5% pregnancy rate. The implementation of all young oncological patient sperm cryopreservation has an important place in our laboratory methods.

6. Trends and future perspectives

Many researchers are studying different methods to improve cryopreservation outcome by modification of essential factors (cryoprotectants, freezing rate, warming). Trends and new perspectivities in cryopreservation in human-assisted reproduction are an important part of this chapter.

6.1. Optimization of current methods

6.1.1. Inhibition of ROCK kinase

Several new steps and procedures for optimization of current methods were developed during last years. It is well known that vitrification procedure often increases apoptosis in embryonic cells, and it results in decrease of developmental competence. Specific postvitrification treatment can suppress this effect in somatic cells or animal oocytes. It was reported that inhibition of Rho-associated coiled-coil kinase (ROCK) improves developmental competence of vitrified/thawed bovine oocytes [22]. This treatment was also effective in human embryonic cells [23], or bovine blastocysts [24]. ROCK kinase is involved in regulation of metabolism, apoptosis, growth, cytoskeletal assembly and cell contraction.

6.1.2. New vitrification devices

New way to increase the cooling rate is reducing the use of cryoprotectants consist in the reduction of liquid nitrogen temperature. In order to avoid vaporization of liquid nitrogen, the temperature is reduced until −210°C, applying a negative pressure [25]. In this condition, nitrogen partially solidifies and creating nitrogen slush, which is less likely to evaporate on contact with specimen compared to liquid nitrogen. This method was very effective in human blastocyst [26]. Cells immersed into nitrogen slush cool more rapidly because they come into contact with liquid nitrogen sooner than those immersed in normal liquid nitrogen. It can provide very high cooling rate (up to 135,000°C/min. The cooling rate is enhanced mainly in the first part of cooling (from 20 to −10°C).

6.1.3. Hydrostatic pressure

Survival of cryopreserved oocytes and embryos is affected by many factors, and their role is still unclear. Recent studies also reported promising results after applying of high hydrostatic pressure during pretreatment of oocytes and embryos. Some studies show that cultivation medium has a dramatic effect on efficiency of cryopreservation methods. However, it was tested that short time exposition of high hydrostatic pressure prior to vitrification (probably thought production of HSP proteins [27]) significantly improved the survival and hatching rate in murine blastocyst [28].

6.1.4. Antioxidative treatment

Oxidative stress has been implicated in many different types of cell injuries, including membrane peroxidation, oxidation of amino acids or nucleic acids, apoptosis and necrosis, which decrease survival rate after cryopreservation. Experiment realized in model animals indicated positive effect of the presence of antioxidant in cultivation medium after thawing of embryos [29]. Indeed, supplementation of α-tocopherol in recovery culture medium resulted in a significantly higher blastocyst yield from the postwarm bovine oocytes in comparison with control oocytes [30]. Actual methods are capable of achieving proper vitrification attaining high level of viscosity and dehydration and delivering high freezing and warming rates. Recent studies realized on experimental animals bring new applicable knowledge suitable for optimization of current method. In our opinion, further research in vitrification media and devices is important for next development of these methods.

It is well known that type of culture media (where are embryos after thawing) is very important for successful thawing process. This fact is often ignored and we believe that gentle appropriate treatment after thawing can improve the survivability of oocytes and embryos.

6.2. New trends

6.2.1. Freeze all

One of the new strategies is also "freeze all". In "freeze all cycle", all embryos are cryopreserved and later used after thawing in another reproductive cycle. This approach is very often used in cooperation with preimplantation genetic screening (PGS) of embryos before their transfer into the uterus. New trend in this approach is genetic screening of blastomere after biopsy at the fifth day of *in vitro* cultivation or later. There is no other way and all embryos must be cryopreserved and stored in liquid nitrogen. Embryos are thawed after final decision about embryo aneuploidy and their suitability for transfer into the uterus.

It was presented that implantation, clinical and ongoing pregnancy rates of ART cycles may be improved by performing cryoembryotransfer compared with fresh embryo transfer [31]. It can be explained by a better embryo endometrium synchrony achieved with endometrium preparation cycles. In frozen embryo transfers, endometrium priming may be achieved with the use of E_2 and P, and the endometrial development can be controlled more precisely than in cycles with gonadotropins [32].

6.2.2. Social freezing

Frozen oocyte replacement is a technique where oocytes are retrieved, frozen, stored and fertilized only after thawing them for transfer. This technique helps women to preserve the future ability of having genetically related children at later point in life. It was first used for cancer patients before chemotherapy or radiotherapy. However, it can be also used for delaying motherhood for any reason, such as an absence of suitable partner or a work career. Large companies like Facebook or Apple have recently included social freezing for female employees as an employment benefit. Indeed, just as for fresh oocyte, the outcome of IVF with vitrified oocytes is highly dependent on maternal age. The most appropriate age for effective cryopreservation is unknown, but ideally, it would be in the early to mid-30s, before age at which woman's fertility naturally declines. Younger women have higher chance that they will never require these eggs. Elder women can be under risk of insufficient procedure with few amounts of oocytes, aneuploidy oocytes and very low probability of pregnancy [33].

7. Conclusion

Finally, it is well known that the embryologist training would have a major bearing on the vitrification outcome. Further vitrification procedural improvements using postvitrification chemical treatment would reduce the high sensitivity of oocytes and embryos to cryopreservation and provide valuable information during an advanced postcryopreservation thawing procedure. Furthermore, the strategic placement of embryonic culture media is very important for a successful freezing/thawing process. This fact is often ignored, and we have determined that gentle appropriate treatment after thawing can improve survivability of human oocytes and embryos in much the same manner as witch model animals.

Acknowledgements

Our work is supported by a grant MH CZ—DRO (FNBr, 65269705), and funds are received from the Faculty of Medicine, Masaryk University Brno, Czech Republic to junior researcher Michal Jeseta.

Author details

Jeseta Michal[1*], Zakova Jana[1], Ventruba Pavel[1], Bartosz Kempisty[2,3,4] and Crha Igor[1]

*Address all correspondence to: jeseta@gmail.com

1 Department of Obstetrics and Gynaecology, Center of Assisted Reproduction, University Hospital Brno and Faculty of Medicine, Masaryk University, Brno, Czech Republic

2 Department of Histology and Embryology, Poznan University of Medical Sciences, Poznan, Poland

3 Department of Anatomy, Poznan University of Medical Sciences, Poznan, Poland

4 Faculty of Public Health, Mieszko I School of Pedagogy and Administartion in Poznan, Poznan, Poland

References

[1] Trounson A, Mohr L. Human pregnancy following cryopreservation, thawing and transfer of an eight-cell embryo. Nature. 1983; 305(5936): 707–709.

[2] Michelmann HW, Nayudu P. Cryopreservation of human embryos. Cell Tissue Banking. 2006; 7: 135–141.

[3] Nagashima H, Kashiwazaki N, Ashman RJ, Grupen CG, Seamark MF, Nottle MB. Removal of cytoplasmic lipid enhances the tolerance of porcine embryos to chilling. Biol Reprod. 1994; 51(4): 618–622.

[4] Ghetler Y, Skutelsky E, Ben Nun I, Ben Dor L, Amihai D, Shalqi R. Human oocyte cryopreservation and the fate of cortical granules. Fertil Steril. 2006; 86(1):210–216.

[5] Stachecki JJ, Wiladsen SS. Spindle organization after cryopreservation of mouse, human and bovine oocytes. Reprod Biomed Online. 2004; 8(6): 664–672.

[6] Konc J, Kanyó K, Kriston R, Somoskői B, Cseh S. Cryopreservation of embryos and oocytes in human assisted reproduction. Biomed Res Int 2014; 2014: ID307268, 1–9.

[7] Liebermann J, Nawroth F, Isachenko V, Isachenko E, Rahimi G, Trucker MJ. Potential importance of vitrification in reproductive medicine. Biol Reprod. 2002; 67(6): 1671–1670.

[8] Jin B, Higashiyama R, Nakata Y, Yonezawa J, Xu S, Miyake M, Takahashi S, Kikuchi K, Yazawa K, Mizobuchi S, Niimi S, Kitayama M, Koshimoto C, Matsukawa K, Kasai M, Edashige K. Rapid movement of water and cryoprotectants in pig expanded blastocysts via channel processes: its relevance to their higher tolerance to cryopreservation. Biol Reprod. 2013; 89: 87.

[9] Pegg DE. The role of vitrification techniques of cryopreservation in reproductive medicine. Hum Fert. 2005; 8: 231–239.

[10] Vajta G. Are programmable freezers still needed in the embryo laboratory? Review on vitrification. Reprod BioMed Online. 2006; 6: 779–796.

[11] Fadini R, Brambillasca F, Renzini M, Merola M, Comi R, De Ponti E, Dal Canto M. Human oocyte cryopreservation: comparison between slow and ultrarapid methods. Reprod BioMed Online. 2009; 19(2): 171–180.

[12] Žáková J, Lousová E, Ventruba P, Crha I, Pochopová H, Vinklárková J, Tesařová E, Nussir M.: Sperm cryopreservation before testicular cancer treatment and its subsequent utilization for the treatment of infertility. Sci World J 2014; 2014: ID575978, 1–5.

[13] Coticchio G, Bromfield JJ, Sciajno R, Gambardella A, Scaravelli G, Borini A, Albertini DF. Vitrification may increase the rate of chromosome misalignment in the metaphase II spindle of human mature oocytes. Reprod BioMed Online. 2009; 19(3): 29–34.

[14] Saragusty J, Arav A. Current progress in oocyte and embryo cryopreservation by slow freezing and vitrification. Reproduction. 2011; 141: 1–19.

[15] Sirard MA. Factors affecting oocyte and embryo transcriptomes. Reprod Domest Anim. 2012; 4: 148–155.

[16] Wang H, Racowsky C, Combelles CMH. Is it best to cryopreserve human cumulus-free immature oocytes before or after in *vitro* maturation? Cryobiology. 2012; 65(2):79–87.

[17] Huser M, Crha I, Hudecek R, Ventruba P, Zakova J, Smardova L, Kral Z. Ovarian tissue cryopreservation—new opportunity to preserve fertility in female cancer patients. Eur J Gynaec Oncol. 2007; 28(4): 249–256.

[18] Donnez J, Dolmans MM, Demylle D, Jadoul P, Pirard C, Squifflet J, Martinez-Madrid B, Van Langendonckt A. Livebirth after orthotopic transplantation of cryopreserved ovarian tissue. Lancet. 2004; 364 (9443): 1405–1410.

[19] Zakova J, Sedlackova M, Polak S, Dumkova J, Ventruba P, Crha I. Methods for preserving fertility in young women suffering from cancer: some aspects of ovarian tissue cryopreservation. Bratisl Lek Listy. 2012; 113(3): 192–194.

[20] Schroder AK, Banz C, Katalinic A, Al-Hasani S, Weiss JM, Dietrich K, Ludwig M. Counselling on cryopreservation of pronucleated oocytes. Reprod BioMed Online. 2003; 6(1):69–74.

[21] Li L, Zhang X, Zhao L, Xia X, Wang W. Comparison of DNA apoptosis in mouse and human blastocysts after vitrification and slow freezing. Mol Reprod Dev. 2012; 79: 229–236.

[22] Hwang IS, Hara H, Chung HJ, Hirabayashi M, Hochi S. Rescue of vitrified-warmed bovine oocytes with rho-associated coiled-coil kinase inhibitor. Biol Reprod. 2013; 89(2): 26.

[23] Li X, Krawetz R, Liu S, Meng G, Rancourt DE. ROCK inhibitor improves survival of cryopreserved serum/feeder-free single human embryonic stem cells. Hum Reprod. 2009; 24: 580–589.

[24] Hochi S, Abdalla H, Hara H, Shimoda M, Morita H, Kuwayama M, Hirabayashi M. Stimulatory effect of Rho-associated coiled-coil kinase (ROCK) inhibitor on revivability of *in vitro*-produced bovine blastocysts after vitrification. Theriogenology. 2010; 73: 1139–1145.

[25] Arav A, Yavi S, Zeron Y, Natan D, Dekel I, Gacitua H. New trends in gamete's cryopreservation. Mol Cell Endocrin. 2002; 187(1–2): 77–81.

[26] Huang ChCh, Lee TH, Chen SU, Chen HH, Cheng TCh, Liu ChH, Yang YS, Lee MS. Successful pregnancy following blastocyst cryopreservation using super-cooling ultra-rapid vitrification. Hum Reprod. 2005; 20(1): 122–128.

[27] Kaarniranta K, Elo M, Sironen R, Lammi MJ, Goldring MB, Eriksson JE, Sistonen L, Hekminen HJ. Hsp70 accumulation in chondrocystic cell exposed to high continuous hydrostatic pressure coincides with mRNA stabilization rather than transcriptional activation. PNAS. 1998; 95: 2319–2324.

[28] Pribenszky C, Molnar M, Cseh S, Solti S. Improving post-thaw survival of cryopreserved mouse blastocyst by hydrostatic pressure challenge. Anim Reprod Sci. 2005; 87: 143–150.

[29] Houseini SM, Forouzanfar M, Hajian M, Asgari V, Abedi P, Hosseini L, Ostadhosseini S, Moulavi F, Safahani Langrroodi M, Sadeghi H, Bahramian H, Eghbalsaied Sh, Nasr-Esfahani MH. Antioxidant supplementation of culture medium embryo development

and/or after vitrification-warming; which is the most important? J Assist Reprod Genet. 2009; 26: 355–364.

[30] Hwang IS, Hochi S. Recent progress in cryopreservation of bovine oocytes. BioMed Res Int. 2014; Article ID 570647.

[31] Shapiro BS, Daneshmand ST, Restrepo H, Garner FC, Aguirre M, Hudson C. Matched-cohort comparison of single-embryo transfers in fresh and frozen-thawed embryo transfer cycles. Fertil Steril. 2012; 99: 389–392.

[32] Roque M, Lattes K, Serra S, Sola I, Geber S, Carreras R, Checa MA. Fresh embryo transfer versus frozen embryo transfer in *in vitro* fertilization cycles: a systematic review and meta-analysis. Fertil Steril. 2013; 99(1): 156–162.

[33] Schattman GL. Cryopreservation of oocytes. N Eng J Med. 2015; 373:1755–1760.

Chapter 11

Quality Control Factors Influencing the Successful and Reliable Implementation of Oocyte and Embryo Vitrification

Mitchel C. Schiewe

Additional information is available at the end of the chapter

http://dx.doi.org/10.5772/65332

Abstract

Clinical vitrification evolved slowly, with interests and acceptance being commercially driven by the development of unique devices, safer solutions, and the misconception that ultra-rapid cooling in an "open" system was a necessity to optimizing vitrification success. Furthermore, the dogma surrounding the importance of cooling rates has led to unsafe practices subject to excessive technical variation and risky modifications to create closed-storage devices. The aim of this chapter is to highlight important quality control factors (e.g., ease of use, repeatability, reliability, labeling security, and cryostorage safety) into the selection process of which device/solution to use, independent of commercial manipulations. In addition, we provide clinical and experimental evidence in support of warming rates being the most important factor determining vitrification survival. Lastly, we exhibit indisputable support that aseptic, closed vitrification systems, specifically microSecure vitrification (μS-VTF), can achieve success with attention to quality control details often lacking in open vitrification devices.

Keywords: blastocyst, cryopreservation, device type, oocyte, quality control, vitrification

1. Introduction

The early successes of William F. Rall, PhD, and coworkers with mammalian embryo vitrification (VTF) were based on extensive experimentation, meticulous solution and straw-handling preparations, and precise straw sealing [1, 2]. Although there was less overt cellular damage, these early investigations simply proved that vitrification was a potentially effective alternative cryopreservation procedure, but not necessarily more effective than conventional slow-freezing methodologies. The degeneration experienced with visually intact vitrified embryos could have been due to the potential cryotoxicity of high-molarity vitrification solutions (e.g., VS3a = 6.5 M glycerol) [3]. An alternative consideration involved the importance of warming rates to prevent recrystallization events that could adversely effect cellular survival of vitrified blastomeres [4]. In the early to mid-1990s, most investigations focused on developing safer, less toxic solutions [5–7] to improve vitrification success. It was widely accepted that the combined use of less concentrated permeating cryoprotective agents (CPAs) made for safer vitrification solutions [6]. Indeed, by combining permeating CPAs (e.g., dimethyl sulfoxide (DMSO), ethylene glycol (EG), and glycerol (GLYC)), and adding other nonpermeating CPAs (e.g., sucrose and ficoll) to create moderately concentrated vitrification solutions, brief intervals of exposure proved to be safe to embryonic blastomeres and oocytes. Combined with the commercial development of novel vitrification devices [8], and the proposed relative importance of cooling rate to vitrification success [9–12], vitrification technology has essentially replaced slow-freezing procedures for human oocytes and embryos in the twenty-first century.

The history and general discussion of vitrification's application to human oocytes, zygotes, and embryos have been previously reviewed [13]. Cryopreservation in the absence of damaging ice-crystal formation (i.e., vitrification) efficiently preserves cell membrane integrity, typically yielding high complete survival rates (>90%) for oocytes and embryos utilizing various vitrification methods. Live birth rates associated with the vitrified embryo transfer cycles are considered equal to or higher than those of fresh blastocysts [14–16], and others claim that the use of vitrified donor oocytes is comparable to fresh donor oocytes [17, 18]. Yet, vitrification success is susceptible to procedural variation between programs referred to as "technical signature" [19]. Variation associated with technical repeatability and reliability between individuals, and a multitude of vitrification devices and methods have resulted in inconsistencies between programs applying vitrification. To optimize the application of vitrification industry-wide, several quality control factors should be taken into account. This chapter describes those quality control factors and problematic events/examples associated with the development of different vitrification devices. In addition, we detail the successful implementation of a noncommercial, simple, and secure aseptic-closed procedure (i.e., microSecure-VTF) which has aimed to minimize quality control-related variation. Furthermore, there is a growing need to educate reproductive biologists about the complexity of the vitrification process and understand the relative importance of warming rate to cooling rate and their relationship to the vitrification solution used.

2. Quality control considerations

In the last decade, vitrification technology has rapidly supplanted conventional freezing practices. To a great extent, this was due to the commercial industry's development of specialized vitrification devices. However, the overall safety, efficiency, and effectiveness of clinical vitrification have been handicapped by these same commercial influences introducing inherent design flaws in devices used in the in vitro fertilization (IVF) industry. Indeed, specific differences in devices and their utilization have introduced significant technical variation between individuals and programs. Although the efficacy of any single vitrification method/device can be optimized within a program (i.e., intra-program variation), through extensive training and experience, its adoption throughout the assisted reproductive technology (ART) industry may be less effective (i.e., inter-program variability). It is this inability to easily and successfully apply a vitrification method between laboratories (i.e., technical signature) that warrants attention, if an optimized universal method(s) is to be executed throughout the IVF industry. When attempting to integrate an effective vitrification system into your clinical laboratory, several quality control factors should be taken into account to fully assess the completeness of a vitrification device and its potential pitfalls. These quality control considerations include the following:

1. **Labeling potential**

   a. Can labels be securely adhered and easily identified?

   b. Do they offer dual color identification potential?

   c. Are they tamperproof and fail safe?

   d. Does it require a secondary label and can the label be easily removed for record-keeping purposes (i.e., patient verification) postwarming?

2. **Technical ease and reliability**

   a. Can embryos be easily loaded into or onto the device in a timely and repeatable manner?

   b. Can the device be easily and safely extracted to facilitate rapid warming (i.e., achieve a warming rate >its cooling rate)?

   c. Can embryos be simply identified and tracked postwarming?

3. **Procedural simplicity and repeatability**

   a. Does the vitrification method offer simplicity and reliability?

   b. Does it easily allow for repeatable applications within and between patients which minimizes variation between technicians (internal) and programs (external)?

4. **LN_2 storage capacity**

   a. Can the devices be easily and safely handled and identified?

   b. Is the device's storage potential space efficient?

   c. Does the device offer security and safety from physical damage?

   d. Does the container provide safety and reliability from possible pathogenic contaminants as an aseptic closed system?

5. **Recovery potential/survivability**
   a. Is the device design prone to potential problems in the guaranteed recovery of embryos?
   b. Will the system reliably vitrify and maintain complete cellular integrity postwarming?

2.1. Vitrification device development

New concepts in vitrification device/container design began emerging in the assisted reproductive technology industry the late 1990s through mid-2000s, as previously reviewed and discussed by Vajta and Nagy [8]. Dr. Vajta and his coworkers created an "open-pulled straw" (OPS) which tapered from the conventional 0.25-ml straw diameter to a >50% reduction in diameter over an approximate 4-cm length [20]. This novel design effectively increased the surface-to-volume ratio which increased its cooling rate capacity in a lower volume, which reduced cryoinjury to vitrified bovine oocytes. The OPS was simple to use (i.e., load) for animal scientists and veterinarians familiar with handling and sealing 0.25-ml straws, yet it was difficult to label and store in a secure, organized, and effective manner. The labeling of the plastic straw with a fine sharpie or cryomarker was subject to being partially rubbed off in cryostorage and becoming un-identifiable. Then, there was also concerns on how to safely and securely store these opened OPS units. One good alternative was to enclose and seal them inside a larger 0.5-ml semen straw [21], preferably an ionomeric resin CBS straw capable of achieving reliable weld seals. Thus, the former quality control issues were resolved at the expense of the insulated OPS having slower cooling rates without direct contact to liquid nitrogen (LN_2). A couple of years later, another adapted straw procedure emerged, called the hemi-straw, which involved supercooling a microdroplet on the inner edge of a transverse cut 0.25-ml straw plunged into LN_2 [22, 23]. Like the OPS, the hemi-straw could be inserted into a 0.5-ml semen straw before LN_2 exposure (as a closed system) or following ultra-rapid cooling and then plug-sealed [24]. The hemi-straw concept led to the commercial development of the aseptic-closed high-security vitrification device (HSV; CryoBioSystem, France) which used a plastic wand device with an elongated trough tip (i.e., gutter) to support a vitrified microdroplet, which was then inserted into a 0.25-ml CBS™ ionomeric resin straw [25].

In the years between the development of the OPS and hemi-straw, a unique carrier system called the "Cryoloop" was adapted from X-ray crystallography applications (Hamilton Research Instruments, USA). A nylon loop, barely detectable to the eye, supported the suspension of a thin film of vitrification solution to facilitate the supercooling (>10,000°C/min) of an embryo or oocyte in the confines of a liquid nitrogen-filled cryovial [26]. Although cryovial labeling and storage were standard practices, the precise loading of the fluid-embryo combination onto and handling of the delicate loop affixed to a metal post was subject to technical variation. Despite the cryoloop's clinical success over the next decade [27, 28], the device required assembly (i.e., glue adhesion of the loop/post to the cap) and the use of specialized instruments (e.g., curved grasping forceps, an extended rod with magnet) to facilitate handling. Although a published comparison between the hemi-straw method and the cryoloop ultimately revealed no differences in survival rates or pregnancy outcomes [24], the proposed importance of ultra-rapid cooling rates to insuring high vitrification success rates dominated the commercial push to integrate vitrification into the human IVF industry. While other novel thin-film, supercooling procedures were proven effective (e.g., electron microscopy (EM) grid, nylon mesh; see review [8]), it was the development of an open-system, plastic wand-flat-blade device called the "Cryotop" [29, 30] that would have the greatest impact on the adoption of clinical vitrification (Kitizato, Japan), as a routine cryopreservation method used for human embryos and oocytes [31]. Promoting the importance of ultra-rapid cooling rates in a micro-volume (0.1 μl), the popularity of open-blade methods grew (e.g., Cryoleaf, Cryolock, and Cryotech) and, like the Cryotop, each device was subject to technical variation and other unique quality control issues discussed below.

While open-system advocates minimized concerns over the potential risks of pathogen cross-contamination among LN_2-stored samples [32], there are others who express strong apprehension over the long-term cryostorage of embryos/oocytes in containers or devices which are unsealed (i.e., leaky, open container, and protected device systems) or poorly/improperly sealed due to disease transmission risk assessments, based on animal model research [33]. Although LN_2 vapor-phase storage tanks offset these concerns, they were and still are not common to, nor practical, in most clinical IVF laboratory settings. Thus, there were additional commercial efforts to produce effective closed vitrification systems in the mid-2000s. During the development of the CBS™ HSV device (mentioned above), an ultra-fine OPS system, called the Cryotip™, was marketed by Irvine Scientific as the first Food and Drug Administration (FDA)-approved vitrification device. This modified closed micropipette system produced comparable postwarming embryo outcomes compared to the Cryotop [30]. Unfortunately, the Good Manufacturing Practice (GMP) focus by the FDA was strictly on the effectiveness of the device to achieve a reliable seal, and not on other important quality control issues influencing gamete and embryo safety. Indeed, the Cryotip™ was a mass-marketed flawed vitrification device that proved to be technically challenging to use (i.e., "technical signature" concept applied) due to aspiration, bubbling, and sealing issues, as well as biosecurity and cryostorage identification issues. Shortly thereafter, another closed micropipette device called the Cryopette™ (Mid-Atlantic Instruments-Origio, USA) was developed to overcome loading and dual-sealing problems associated with the Cryotip™. In addition, this device added color coding, a positive feature originally found in CBS™ 0.3-ml embryo straws. By mounting a colorized, cryo-resistant bulb on one end of a shortened flexipette, it strived to control technical aspiration and expulsion of embryos, and simultaneous close one end. Again, FDA's approval of this device insured that the open end of the flexipette could be effectively sealed without harm to its cellular contents, but did not address concerns regarding labeling, cryostorage safety, or bulb reliability. These quality control flaws were left to the consumer to discover, as discussed below.

The Cryotip and Cryopette devices are both considered "closed systems" as the gametes and embryos they contain are sealed in an environment away from any potential contact with liquid nitrogen. However, the outer surface of their micropipettes still reside in direct contact with LN_2, and thus are still at potential risk of being a carrier of pathogens found free floating in stored LN_2 (e.g., adherent bacteria [34, 35]). Although the risk of transfection is unproven, risk assessment potential is virtually eliminated in an "aseptic, closed vitrification system" such as the HSV [25] and enclosed OPS or cut standard straw [36, 37] approaches (described previously). Unlike many original suboptimal designs, CryoBioSystems made improvements in their HSV system enhancing the ease and reliability of device extraction to reduce warming variation, as well as improving device identification by offering color coding. Subsequently, a similar device, referred to as Vitrisafe (Astro-med Tec, Austria), has also produced a high level of clinical success [38, 39], similar to open systems. In the last decade, three additional novel approaches were developed, with two in particular, the rapid-i [40] and microSecure vitrification [41] being clinically validated. The rapid-i (Vitrolife, USA/Sweden) is a hybrid-designed device mimicking both the flat-blade wand of a cryotop possessing a micro-hole drilled in the center of the surface to suspend the vitrification solutions, like the Cryoloop. The advantage of the hole in the plastic blade was the ability to directly view the embryo in a 0.05-μl volume, with residual vitrification solution easily aspirated from the blade surface. However, technical precision is still required in terms of embryo/oocyte handling, but with less concern aspirating residual fluid off the blade. The rapid-i system has a special LN_2 bath that allows closed bottom-weighted straws to be supported upright in LN_2 with the open end being accessible above a covered surface. Each rapid-i wand could then be supercooled inside each straw, theoretically in a rapid manner, followed by heat sealing and LN_2 storage. Unfortunately, one unexpected problem was the latent conversion of LN_2 vapor to liquid inside the straw during the equilibration period, resulting in the transfer of kinetic energy to a warm wand dropped into the straw. To prevent the initial expulsion of the device, the company adopted a procedural step to cover the straw opening upon device insertion, followed by sealing. Other than that, the straw does not have any colorized component, or system for secure or duplicate labeling.

The growing high level of success and undeniable security advantages of some aseptic closed systems [38–42] has prompted another new and potentially problematic development of hybridizing vitrification systems. Attempting to gain the benefits of direct LN_2-mediated ultra-rapid cooling, some innovative embryologists and at least one commercial company have begun sealing LN_2-exposed open devices into plastic straws. Unlike the safety and security of weld-sealing an ionomeric plastic straw under ambient (20–22°C) conditions (i.e., HSV and mS-VTF), the compliance of supercooled straws to effective heat sealing may be compromised leading to suboptimal, unsecure closure. Without the specialized LN_2 bath lid of the rapid-i device, the sealing of straws while primarily submerged in LN_2 could lead to the incomplete heat sealing of straws and/or the partial trapping of N_2 gas inside a straw. Upon rapid warming, the consequences of such a scenario could be disastrous, as the rapid expansion of N_2 gas from a liquid phase can be explosive in a closed container [43]. In an at-risk situation, as described above, the straw should be cut and the vitrification device removed while still partially submerged in LN_2. Furthermore, if the warming rate of a hybrid device does not exceed its initial cooling rate, the viability of its vitrified gametes or embryos will be compromised.

2.2. Relative importance of warming rates

Over 60 years ago, Dr. Peter Mazur first discussed the relative importance of warming rates to cellular solutions which had been cooled very rapidly [44]. He proposed that unstable ice crystals can grow to a size damaging cells if the warming velocity was not sufficiently high to melt the unstable ice formation. Then, with the development of vitrification in the 1980s, both Drs. Greg Fahy [45] and William Rall [4] warned us about the importance of warming for cellular survival. Yet, it was Dr. Mazur again, with his postdoctoral fellows and scientific colleagues, who defined our path to successful vitrification over the past decade. The answer today is definitive, and the efficacy of vitrification success is more highly dependent on warming rates than cooling rates [46–50]. Independent of the vitrification device or open/closed system used, the warming rate must exceed the cooling rate to insure high survival rates. Using a slow warming model, Dr. Brian Wowk has demonstrated the relationship of ice nucleation during cooling and recrystallization of ice growth upon warming relative to cryoprotective agent concentration [50], as well as the thermodynamics behind vitrification [51]. Under low-warming conditions, today's typical commercial vitrification solutions (e.g., 30–32% (total permeable cryoprotective agents), EG/DMSO or EG/PPG) are classified as "unstable" and are highly dependent on rapid cooling and higher warming rates for cell survival. Whereas metastable solutions (e.g., 50–70% (total permeable cryoprotective agents)) have a lower temperature of heterogeneous ice nucleation (T_h) where the warming rate does not need to outrun the temperature of devitrification (T_d) to inhibit (i.e., melt) the potentially damaging recrystallization of ice, as originally eluded to by Mazur [44].

Although commercial vitrification solutions work well in both open and closed systems, the use of metastable solutions (e.g., VS3a, 6.5 M glycerol or ICE-BL, >7.9 M glycerol/EG) may offer aseptic closed vitrification systems a higher level of biosafety. As we learned in the 1990s, the mixture of cryoprotective agents reduces the cytotoxicity for a potential vitrification solution [52]. Additional research from Mazur's laboratory [53] has shown that infrared laser technology can be used to exponentially increase warming rates and achieve high oocyte survival using a threefold diluted vitrification solution. But is there really a need to make solutions even less concentrated at the expense of becoming warming rate dependent? Concerns over the potential toxicity of vitrification solutions are likely as misunderstood, as the importance of cooling rates to successful vitrification. Recently, we have shown that human blastocysts (BL) diluted into a more concentrated ICE-BL non-DMSO vitrification solution (Innovative Cryo Enterprises, USA) are as viable as those in 30% EG/DMSO (LifeGlobal, USA/Canada) and 32% EG/PPG (Vitrolife, USA/Sweden) solutions for up to 10-min exposure [54], revealing that human blastocysts are more resilient to vitrification toxicity than previously believed. Furthermore, we conducted a series of revitrification (rVTF) studies aimed at understanding the cryotoxicity and osmotic stress associated with different vitrification solutions [54]. Using our control metastable vitrification solution (ICE-BL), no difference in 0-h survival or 24-h development was exhibited after up to 5X rVTF, with or without sucrose elution between treatments, proving how cryotolerant human blastocysts are to metastable vitrification in an aseptic closed system. Interestingly, ongoing unpublished data using a common EG/DMSO solution revealed no difference in survival, but a significant decline in sustained viability at 24 hr after the second rVTF treatment. Like Wowk's slow-warming model, our data may be revealing the vulnerability of unstable solutions to cryoinjury when exposed to a cryostress model. These studies demonstrate interesting findings in support of theoretical vitrification principles regarding the relationships of cooling and warming rates relative to molar concentration of cryoprotective agents. In addition, it reveals that the commercial industry should seriously reevaluate vitrification formulations to optimize their product for the IVF industry [55].

2.3. Identifying and troubleshooting device-related quality control problems

A huge problem among early non-straw or cryovial products was in how a particular device was labeled. Most experienced embryologists with good laboratory practices found effective ways to properly label a device (as described below in Section 3). When using Brady labels to wrap around an open-system handle or a 0.25-ml straws, care is needed to insure the font size is readable. A horizontal wrap may be more secure than a vertical placement, but will likely not provide sufficient space for readable text. Therefore, validation testing should be performed to confirm that the vertical surface is reliably adherent. Alternatively, the label could be adhered horizontally to create an external flag on the device, which optimizes the labeling surface. If a flag label is used, be cautious to not overcrowd samples causing possible breakage of the flag. In most cases, a secondary ID on the device is warranted to prevent possible identity loss in storage. For example, the Cryotip was a poorly planned device in terms of labeling and storage, but there was a simple solution suggested to the company after their FDA approval and marketing commenced. The user had the option of placing or handwriting (with sharpie) the label onto the metal protector or handwriting of the upper straw by retracting the metal protector in a sterile manner during setup. Difficulty in the ease and safety of extracting these miniature devices from a shortened goblet for identification/selection was a commonly experienced storage problem (discussed further below). The simple solution was the use of 0.5-ml straws crimped at the open end and slid over the sealed end of the device, as used for conventional one-step straws. Each straw handle could be used to correspondingly label each device to facilitate safe and easy identification under ambient conditions while the device remained safely submerged in LN_2. Such a solution was not possible to the alternative Cryopette system, which may have improved the system for aspirating embryos into the device but did so at the expense of the important factor of labeling. Perhaps, the most outlandish experience I witnessed was with a shipment of OPS units (n = 8–10 OPS) simply stored in a flat cartridge meant to hold straws. Each OPS only had a hand-printed last name and a date, without any further identification distinguishing them. Then to make matters worse, when I attempted to slide the wand upward to systematically extract them, they did not move up like a straw but instead the tapered tips slid down and jammed up the glide track creating a real problem. These possible conditions must be carefully thought out before implementation.

Technical ease and reliability of the methods used are an important consideration. This factor is important within your group, but perhaps even more important outside your program. Often times patients move or simply change physicians (i.e., clinics/laboratories), resulting in the transport of your embryos and oocytes to another laboratory. To avoid possible liability issues associated with lost embryos, faulty devices, or nonsurvival issues of the patients' only/last embryo, it is critical to insure the end user will also be successful. Thus, simple and reliable products are essential. Early open systems such as EM grids and nylon mesh units were difficult to use for an individual unfamiliar with the device. Even a more established product

http://dx.doi.org/10.5772/65332

such as Cryoloops presented challenges to the unknowing user. For example, upon warming if the supercooled metal post holding the loop, containing the embryo(s), contacted the warming medium it would cause excessive vaporization and bubbling which would hinder an efficient recovery process. Worst yet, the Cryoloop device required assembly (i.e., glue adhesion of the loop/post to the cap) and was susceptible to device error associated with loose/fallen posts, broken loops, as well as variable microdroplet sizes. The Cryotip has also been known to be susceptible to recovery problems associated to excessive internal bubbling and damaged tips. Then, there are the popular open-blade methods that predominate the world-wide ART industry. Tremendous technical variation exists with respect to the amount of vitrification solution to retain on the surface with the embryo or oocyte(s). If too large, the droplet could disengage from the surface during storage, or in a closed system such as the HSV the droplet could displace itself to the inner straw surface. At least in the closed system, there is still the opportunity to recover the lost embryo or oocytes from the sterile inner straw surface. If the microdroplet surrounding the embryo/oocyte(s) is aspirated too much (i.e., nearly dry), it places the embryo/oocyte(s) at risk of dehydration and osmotic injury prior to vitrification. One final example worth acknowledging here is a more recently developed device called the Cryotech, made of a lighter weight, more transparent film with a 90° angle to the embryo/oocyte-loading surface. On at least three occasions, involving the international shipment of oocytes, the oocytes were lost upon warming. The last shipment was actually tracked by the same experienced, senior embryologist performing both vitrification and warming events in two different countries. In the latter situation, the device failed, suggesting that the excessive handling dislodged the droplet from the surface. These are the types of very unpleasant circumstances that typically leave each party, or worse the patient, blaming negligence of one laboratory or the other, but could have simply been the fault of the device design. In these scenarios, a closed pipetting device, such as microSecure-vitrification (μS-VTF), is ideal in its simplicity and reliability to retain the cellular products they contain.

Emphasis on procedural simplicity and repeatability cannot be underestimated. Ideally, you want to minimize the time and effort required to place the embryo/oocyte(s) onto or into a device. We have already seen problematic examples above that can create significant hardship for all parties involved. Thus, it is critical that we strive to use fail-safe systems, and that the procedures involved are also safe for the embryologist to perform. An important technical example here is the sealing or securing of a device once the embryo/oocyte(s) are loaded onto the device. Most closed systems are loaded at room temperature and their straws heat sealed, preferably in an ionomeric-resin straw (e.g., CBS™) and using an automatic sealer. However, if using a manual impulse sealer or other miscellaneous approaches (e.g., heated forceps, curling iron, etc.) then a meticulous quality control practice must be implemented to insure the completeness of each straw seal by each technician. One approach I learned from Dr. Rall over 30 years ago on the sealing of conventional 0.25-ml plastic straws with a standard impulse sealer was to flip them 180° several times until it adheres to the Teflon surface (over the electrode) and then do it one more time (requiring a slight delay to gently pry the straw from the cooling surface). That technique was repeatable and teachable, and more importantly created reliable and secure seals that never resulted in an exploding straw post rapid warming. If the seal is incomplete due to a poor sealing technique or a noncompliant plastic due to LN_2

vapor conditions (described above), these straws will allow LN_2 seepage into the container to occur. If these containers are warmed too rapidly, the vaporization pressure could be excessive and damaging. One other noteworthy example here is a device requiring two different heat settings to optimally seal two different size openings (e.g., Cryotips). The latter was simply a formula for repeated errors, frequently resulting in bent and burnt tips. It is also possible that the overheating of the fine tip ends may have been partially responsible for the excessive and problematic bubbling experienced in these devices.

Another problem experienced by many inexperienced user of the Cryotip device was the overcrowding of devices within a given storage goblet. Unlike a compact arrangement of straws, the tight opposition of Cryotips could cause their protective shields to rise, leaving their delicate tips exposed to damage (e.g., bending, breakage). Similarly, although Cryopettes were not as delicate, they were completely unexposed in storage without support. The potential for breakage or fracturing its bulb connection, if accidentally compressed in frozen storage, was a real risk. Another important practical factor to consider is the LN_2 storage capacity of a device. If we consider that 0.25-ml straws (e.g., HSV, Vitrisafe) or perhaps Cryotop devices in large goblets is an optimal standard of 10 units/goblet, then the storage of 8 units of 0.5-ml straw-size devices (e.g., rapid-i, μS-VTF) or square-capped Cryolock devices is very good. However, the safe storage of five Cryotips per goblet begins to become inefficient, while one or two Cryoleaf devices are completely impractical. Lastly, we have already discussed the ability to safely access and visualize samples in storage/LN_2-filled dewar flasks or specialized bathes, but what about the safety of the handler. Most open-system methods require the placement of a protective straw cover (e.g., Cryotop, Cryotech) or a plastic cap (e.g., Cryolock) over the supercooled device end for storage protection. Likewise, these protective covers must be removed in LN_2 prior to warming to facilitate high warming rates. However, these covers can be difficult to unlock and remove under freezing conditions. Any miscues in the insertion or removal of the protective covers could adversely influence the stability of the embryo/oocyte(s) on the surface of the open blade. Both vitrification and warming events entail the coordinated handling of device components, with fingers and forceps, in close proximity to LN_2, thus creating reoccurring safety issues. Although the use of protective liners provides delicate finger agility in handling and reduces potential contact burns, it does eliminate a mishandling event (e.g., connecting or sealing hybrid devices) that could result in the accidental LN_2 spillage of a full dewar flask. In short, the unnecessary handling and manipulation of devices in LN_2 creates biosafety issues for the user. By contrast, the assembly and sealing of aseptic closed devices under ambient conditions eliminates similar safety concerns.

Our final end point consideration is the recovery and survival rate potential of a given device. As we have already discussed above, there are several quality control factors that can ultimately influence the final outcome. One issue we have not touched on is the advantage of being able to visualize the embryo or oocyte(s) upon warming. This is particularly important with oocytes as they do become highly translucent during their initial exposure to the T1 sucrose solution. Therefore, methods that allow you to distinctly image and account for the expected number of embryo(s)/oocyte(s) present (e.g., Cryoloops, rapid-i and pipetting methods) offer distinct advantage to efficiently locate the desired cell products. Blade and hemi-straw microdroplet methods can leave the technician wondering if the unfound embryo(s) and oocyte(s) are still on the device or free floating on the surface of the sucrose solution or attached to an air bubble. The fact is problems can arise and some methods simply make it easier to troubleshoot the issue at hand. Unfortunately, most technical and clinical publications failed to discuss recovery rate potential and associated problems, but instead choose to disguise that outcome among the nonsurvival group. It is unclear why that has been a scientifically acceptable practice, considering rare embryo losses using conventional slow-freezing technology typically warranted an incident report. If we are to fully evaluate the efficacy of a vitrification device or our ability to efficiently apply the technology, we must be willing to honestly share our mistakes and device experiences, as touched on by Vajta and others [56].

3. The microSecure-VTF (μS-VTF): a quality control solution

Having a firm grasp of the cryobiological principles of vitrification, we developed an aseptic closed vitrification device aimed at insuring the simplicity, efficacy, and reliability of vitrification success [57]. It was developed in 2008 as an inexpensive, noncommercial, FDA-compliant method which optimized quality control aspects of vitrification to reduce or eliminate technical variation. Using the CBS™ 0.3-ml embryo straw (with hydrophobic plug) as our model, our system uniquely offers tamperproof internalized, dual-colored labeling. The use of different label and rod colors allows for quick identification of patient samples based on day of cryopreservation, whether blastocyst biopsy was performed, or blastocyst quality, for example. In contrast to the HSV system, we maintained secure labeling by not reducing the straw diameter. Having internalized labels allows us to use nonpermanent adhesion labels (GA International, USA) that can be easily removed postwarming and placed onto the patients' Cryo-data sheet record to confirm identification with the patient at the time of ET. Furthermore, in the case of a preimplantation genetic screening (PGS) cycle with discard aneuploidy embryos, the placement of removed labels onto the Cryo record is an excellent quality assurance practice. Finally, in terms of labeling it is essential that an accurate description of the patient sample is conveyed, including the last and first name, secondary ID, embryo description (#, stage, quality grade; Ex: 1x4AA or 1x8cB), and the cryopreservation date. Upon receiving other devices in our laboratory, it is so surprising to witness how little information some programs actually provide on a device. Out of respect to all IVF laboratories, proper labeling is essential to avoid possible liability issues.

Since μS-VTF uses shorten sterile flexipettes to pipette, load, and directly store embryo(s) or oocytes, there is no secondary device surface to introduce technical variation. Thus, μS-VTF embryos and oocytes are simply loaded and easily visualized upon removal to insure >99.9% recovery rates. To achieve rapid warming after safe patient sample identification in a dewar flask, the straw is cut below the plug/seal (below the ID rod) and quickly tipped (60° angle) and tapped to promote the free fall of the flexipette into a warm sucrose bath (see You Tube video "microSecure vitrification warming"). On rare occasion, if an embryo is missing upon pipetting into T1 solution it has invariably been found in the sucrose bath, due to it having been loaded to close to the tip. From this rare experience, we have learned that although there is capillary drift into flexipettes while resting on the sidewall of a 60-mm dish in the sucrose bath (for 5–10 s, as the pipette fluid volume will attempt to equilibrate to the sucrose level), the initial plunge into the bath may create an initial force that pulls a fraction of fluid from the tip. It is important that biologists remain mindful to load the embryo(s)/oocytes approximately mid-way in the fluid column. Again, we control this by aspirating a full, fresh 3-μl column of vitrification solution into the pipette (i.e., plunger released, no technical variation) and then expel one-third to a half of the fluid upon picking up the embryo(s)/oocytes, followed by controlled plunger release (to preset fill volume). Upon pipette removal and tip drying (i.e., sterile gauze wiping), the capillary volume in the flexipette is stable during handling procedures. Our rare loss of an embryo has been exclusively related to hatched blastocysts post-biopsy. These embryos can be extremely adherent on contact with any plastic (i.e., charged surface) and potentially difficult to ID in their completely collapsed state. Thus, as with our standard blastocyst biopsying of trophectoderm cells, we suggest pre-coating the surface of all pipettes with human serum albumin (HSA) before handling to minimize cellular stickiness and possible loss of hatched blastocysts.

Next, the μS-VTF system uses CBS™ ionomeric-resin straws that completely weld seal using an automated sealer, which again effectively eliminates technical variation. By not worrying about the quality of the seal, our system offers repeatability and reliability only found in CBS™ straw products (e.g., HSV). Prior to sealing, we make sure the tip of the flexipette has dropped down to the plug end, insuring at least 1 cm of air space to safely seal the straw. Next, we suggest supporting the straw at the point of sealer contact (as opposed to the natural instinct to hold the end of the straw) to minimize any abrupt vibration stimulated by the automatic sealer. Upon inverting the straw label-end up, we check the quality of the seals and whether any fluid remnant/discharge appears in the upper straw air space (as the flexipette base should now be resting against the bottom seal). The upper air space near the plug/labeling rod insures safety to cut the straw postwarming. If any fluid was visualized, we check to make the flexipette did not accidentally get sealed into the straw. If on a rare occasion this happened: (1) if the seal is incomplete then you must extract the flexipette and attempt to find the embryo in the residual fluid droplet before reloading; or (2) if the seal is complete, simply make a note on the record (for that straw #) of the situation, so that proper care is taken postwarming to rinse the inner straw for possible extruded oocytes/embryo(s). Upon storing the straws in LN_2 on canes with large open goblets, up to eight straws can be stored/cane (i.e., good storage capacity). Furthermore, there is no need for an upper cover on the cane, as each straw is weighted, unless they are transported and susceptible to not maintaining their upright position. Coincidentally, if a straw is ever to accidentally drop into an LN_2 tank, they are easily recovered as the weighted rod drops the tank bottom and sticks straight upward (due to air buoyancy in the straw), as opposed to lying on the bottom somewhere in the residual N_2 debris.

As an aseptic closed system whose vitrification device (i.e., a sterile flexipette) is stored in an outer straw container, the μS-VTF device achieves a cooling rate of 1391°C/min and corresponding warming rate of over 6000°C/min. As an insulated device with lower cooling rates than an open device system, it has proven to be more resilient to accidental room temperature exposures (Ovation Fertility, unpublished data). Overall, the µS-VTF device has been systematically validated to be a simple and reliable approach that minimizes intra- and inter-laboratory technical variation, while providing maximum cryosecurity using sterile products [41]. In addition, it has been developed without commercial influence and marketing pressure, thus providing the added benefit of substantial cost-savings. In today's IVF industry, which is increasingly reliant of biopsying and vitrifying every fair to excellent quality blastocyst to optimize pregnancy success [58], costs are an increasingly important factor to consider. This is especially true when one realizes that 50–75% of the genetically tested blastocysts will be aneuploidy and destine to be discarded after short-term storage. In conclusion, the µS-VTF system has proven to be a highly effective procedure that may offer "universal" acceptance to alleviate current quality control concerns with the handling, storage, and shipment of vitrified oocytes and embryos.

4. Conclusion

Vitrification is the single most impactful assisted reproductive technology in the IVF industry since the development of intracytoplasmic sperm injection (ICSI). Today, we faithfully cryopreserve blastocysts and oocytes without regard to possible loss. We have had to reeducate ourselves, and our infertility patients, that fresh ET is no longer better than vitrified ET cycles, especially in combination with blastocyst biopsying and preimplantation genetic screening. By adhering to strict quality control standards and quality assurance practices, we can continue to improve the reliability of our laboratory outcomes, and help avoid future liability issues together.

Acknowledgements

I want to express my gratitude to Dr. Peter Mazur, "The Father of Modern Cryobiology," who passed away on December 30, 2015, at the productive age of 88. Dr. Mazur continued mentoring fellows and conducting studies into his final year and for the last decade provided us critically important insights into vitrification, specifically the relative effects of warming on oocyte and embryo survival and viability. As he completed the scientific journey he had started more than 60 years earlier, he was able to clearly demonstrate to us the relative importance of warming over cooling rate. Thank you for all that you taught us that made these technological advancements possible. And finally, thank you for making it so transparent that "The rate of warming must be equal to or exceed that of the cooling rate, if we are to optimize survival." I am so grateful to have had the opportunity to have had a couple of memorable phone conversations with you in 2015. Godspeed Dr. Mazur and rest in peace, knowing that your scientific contributions will educate future generations.

Author details

Mitchel C. Schiewe

Address all correspondence to: mschiewe@ovationfertility.com

Ovation Fertility, Newport Beach, CA, USA

References

[1] Rall WF, Wood MJ, Kirby C, Whittingham DG. Development of mouse embryos cryopreserved by vitrification. J Reprod Fertil. 1987;80:499–504.

[2] Schiewe MC, Rall WF, Stuart LD, Wildt DE. Ovine embryo cryopreservation: Analysis of cryoprotectant, cooling rate and *in situ* straw dilution using conventional freezing or vitrification. Theriogenology. 1991;36:279–93.

[3] Fahy GM. The relevance of cryoprotectant "toxicity" to cryobiology. Cryobiology. 1986;23:1–13.

[4] Rall WF. Factors affecting the survival of mouse embryos cryopreserved by vitrification. Cryobiology. 1987;24:387–402.

[5] Kasai M, Nishimori M, Zhu SE, Sakurai T, Machida T. Survival of mouse morulae vitrified in an ethylene glycol-based solution after exposure to the solution at various temperatures. Biol Reprod. 1992;47:1134–9.

[6] Ali J, Shelton JN. Design of vitrification solutions for the cryopreservation of embryos. J Reprod Fertil. 1993;99:471–7.

[7] Shaw JM, Kuleshova LL, MacFarlane DR, Trounson AO. Vitrification properties of solutions of ethylene glycol in saline containing PVP, Ficoll, or Dextran. Cryobiology. 1997;35:219–29.

[8] Vajta G, Nagy ZP. Are programmable freezers still needed in the embryo laboratory? Review on vitrification. Reprod Biomed Online. 2006;12:779–96.

[9] Lane M, Schoolcraft WB, Gardner D. Vitrification of mouse and human blastocysts using a novel cryoloop container-less technique. Fertil Steril. 1999;72:1073–8.

[10] Kuwayama M, Vajta G, Kato O, Leibo S. Highly efficient vitrification method for cryopreservation of human oocytes. Reprod Biomed Online. 2005a;11:300–8.

[11] Kuwayama M, Vajta G, Ieda S, Kato O. Comparison of open and closed methods for vitrification of human embryos and elimination of potential contamination. Reprod Biomed Online. 2005b;11:608–14.

[12] Kuwayama M. Highly efficient vitrification for the cryopreservation of human oocytes and embryos: The Cryotop method. Theriogenology. 2007;67:73–80.

[13] Liebermann J. In: Katkov I, editor. Frontiers in Cryobiology. 1st ed. InTech, Rijeka: Croatia; 2014. p. 169–84.

[14] Zhu D, Zhang J, Cao S, Zhang J, Heng BC, Huang M, Ling X, Duan T, Tong GQ. Vitrified-warmed blastocyst transfer cycles yield higher pregnancy and implantation rats compared with fresh blastocyst transfer cycles-time for a new embryo transfer strategy? Fertil Steril. 2011;95:1691–5.

[15] Grifo JA, Hodes-Wertz B, Lee H-L, Amperloquio E, Clarke-Williams M, Adler A. Single thawed euploid embryo transfer improves IVF pregnancy, miscarriage, and multiple gestation outcomes and has similar implantation rates as egg donation. J Asst Reprod Genet. 2013;30:259–64.

[16] Whitney JB, Schiewe MC, Anderson RE. Single center validation of routine blastocyst biopsy implementation. J Asst Reprod Genet. 2016 in press; doi:10.1007/s10815-016-0792-3.

[17] Nagy ZP, Chang C-C, Shapiro DB, Bemal DP, Elsner CW, Mitchell-Leef D, Toledo AA, Kort HI. Clinical evaluation of the efficiency of an oocyte donation program using egg cryo-banking. Fertil Steril. 2009;92:520–6.

[18] Cobo A, Meseguer M, Remohi J, Pellicer A. Use of cryo-banked oocytes in an ovum donation programme: A prospective. Randomized, controlled, clinical trial. Hum Reprod. 2011;25:2239–46.

[19] Stachecki JJ, Garrisi J, Sabino S, Caetano JPJ, Wiemer K, Cohen J. A new safe, simple, and successful vitrification method for bovine and human blastocysts. Reprod Biomed Online. 2008;17:360–7.

[20] Vajta G, Holm P, Kuwayama M, Booth PJ, Jacobsen H, Greve T, Callesen H. Open pulled straw (OPS) vitrification: A new way to reduce cryoinjuries in bovine ova and embryos. Mol Reprod Devel. 1998;51:53–8.

[21] Vajta G, Lewis IM, Kuwayama M, Greve T, Callesen H. Sterile application of the open pulled straw (OPS) vitrification method. Cryo Lett. 1998;19:389–92.

[22] Vanderzwalmen P, Bertin G, Debauche C, *et al*. Births after vitrification at morula and blastocyst stages: Effect of artificial reduction of the blastocoelic cavity before vitrification. Hum Reprod. 2002;17:744–51.

[23] Vanderzwalmen PV, Bertin G, Debauche C, Standaer V, Bollen N, van Roosendal E, Vandervorst M, Schoysman R, Zech N. Vitrification of human blastocysts

with the hemi-straw carrier: Application of assisted hatching after thawing. Hum Reprod. 2003;18:1504–11.

[24] Liebermann J, Tucker M. Effect of carrier system on the yield of human oocytes and embryos as assessed by survival and developmental potential after vitrification. Reproduction. 2002;124:483–9.

[25] Van Landuyt L, Verpoest W, Verheyen G, De Vos A, Van de Velde H, et al. Closed blastocyst vitrification of biopsied embryos: Evaluation of 100 consecutive warming cycles. Hum Reprod. 2010;26:316–22.

[26] Lane M, Schoolcraft WB, Gardner D. Vitrification of mouse and human blastocysts using a novel cryoloop container-less technique. Fertil Steril. 1999;72:1073–8.

[27] Mukaida T, Nakamura S, Tomiyama T, Wada S, Kasai M, Takahashi K. Successful birth after transfer of vitrified human blastocysts with use of a cryoloop containerless technique. Fertil Steril. 2001;76:618–23.

[28] Mukaida T, Oka C, Goto T, Takahashi K. Artificial shrinkage of blastocoeles using either a micro-needle or a laser pulse prior to the cooling steps of vitrification improves survival rate and pregnancy outcome of vitrified human blastocysts. Hum Reprod. 2006;21:3246–52.

[29] Katayama KP, Stehlik J, Kuwayama M, Kato O, Stehlik E. High survival rate of vitrified human oocytes results in clinical pregnancy. Fertil Steril. 2003;80:223–4.

[30] Kuwayama M, Vajta G, Leda S, Kato O. Comparison of open and closed methods for vitrification of human embryos and elimination of potential contamination. Reprod Biomed Online. 2005;11:608–14.

[31] Kuwayama M. Highly efficient vitrification for the cryopreservation of human oocytes and embryos: The Cryotop method. Theriogenology. 2007;67:73–80.

[32] Pomeroy KO, Harris S, Conaghan J, Papadakis M, Centola G, et al. Storage of cryopreserved reproductive tissues: Evidence that cross-contamination of infectious agents is a negligible risk. Fertil Steril. 2010;94:1181–8.

[33] Rall W. Avoidance of microbial cross-contamination of cryopreserved gametes, embryos, cells and tissues during storage in liquid nitrogen. Embryologists Newsletter. 2003;6(2):1–7.

[34] Bielanski A, Bergeron H, Lau PC, Devenish J. Microbial contamination of embryos and semen during long term banking in liquid nitrogen. Cryobiology. 2003;46:146–52.

[35] Bielanski A, Vajta G. Risk of contamination of germplasm during cryopreservation and cryobanking in IVF units. Hum Reprod. 2009;24:2457–67.

[36] Isachenko V, Katkov I, Yakovenko S, Lulat A, Ulug M, et al. Vitrification of human laser treated blastocysts within cut standard straws (CSS): Novel aseptic

packaging and reduced concentrations of cryoprotectants. Cryobiology. 2007;54:305–9.

[37] Isachenko V, Montag M, Isachenko E, Zaeva V, Krivokharchenko I, et al. Aseptic technology of vitrification of human pronuclear oocytes using open-pulled straws. Hum Reprod. 2005;20:492–6.

[38] Panagiotidis Y, Vanderzwalmen P, Prapas Y, Kasapi E, Goudakou M, et al. Open versus closed vitrification of blastocysts from an oocyte-donation programme: A prospective randomized study. Reprod Biomed Online. 2013;26;470–6.

[39] Papatheodorou A, Vanderzwalmen P, Panagiotidis Y, Prapas N, Zikopoulos K, et al. Open versus closed oocyte vitrification system: A prospective randomized study. Reprod Biomed Online. 2013;26:595–602.

[40] Hashimoto S, Amo A, Hama S, Ohsumi K, Nakaoka Y, Morimoto Y. A closed system supports the developmental competence of human embryos after vitrification. J Asst Reprod Genet. 2013;30:371–6.

[41] Schiewe MC, Zozula S, Anderson RE, Fahy GM. Validation of microSecure vitrification (μS-VTF) for the effective cryopreservation of human embryos and oocytes. Cryobiology. 2015;71:264–72.

[42] Lopes AS, Frederick V, Van Kerkhoven G, Campo R, Puttemans P, Gordts S. Survival, re-expansion and cell survival of human blastocysts following vitrification and warming using two vitrification systems. J Asst Reprod Genet. 2015;32:83–90.

[43] Schiewe MC, Schiewe E, Vu VN, Zozula S, Anderson RE. Liquid nitrogen vapor sealing of straw containers can be unsafe and detrimental to embryo survival. Austin J Reprod Med. Infertil. 2016;3:1038–41.

[44] Mazur P. Causes of injuries in frozen and thawed cells. Fed Proc. 1965;25(suppl):S175–82.

[45] Fahy GM. Biological effects of vitrification and devitrification. In: Pegg DE, Karow AM Jr., editors. The Biophysics of Organ Cryopreservation. 1st ed. New York: Plenum; 1987. p. 265–93.

[46] Seki S, Mazur P. Effect of warming rate on the survival of vitrified mouse oocytes and on the recrystallization of intracellular ice. Biol Reprod. 2008;79:727–37.

[47] Seki S, Mazur P. The dominance of warming rate over cooling rate in the survival of mouse oocytes subjected to a vitrification procedure. Cryobiology. 2009;59:79–82.

[48] Seki S, Mazur P. Ultra-rapid warming yields high survival of mouse oocytes cooled to –196°C in dilutions of a standard vitrification solution. PLoS One. 2012;7:e36058.

[49] Mazur P, Seki S. Survival of mouse oocytes after being cooled in a vitrification solution to –196°C at 95°C to 70,000°C/min and warmed at 610°C to 118,000°C/min: A new paradigm for cryopreservation by vitrification. Cryobiology. 2011;62:1–7.

[50] Wowk B. Metastable vitrification of cryoprotective solutions. In: Proceedings of the 50th Ann Mtg for Society for Cryobiol. 2013. You Tube video "Cryo2013".

[51] Wowk B. Thermodynamic aspects of vitrification. Cryobiology. 2010;60:11–22.

[52] Ali J, Shelton JN. Design of vitrification solutions for the cryopreservation of embryos. J Reprod Fertil. 1993;99:471–7.

[53] Jin B, Kleinhans FW, Mazur P. Survival rates of mouse oocytes approach 100% after vitrification in 3-fold diluted media and ultra-rapid warming by an IR laser pulse. Cryobiology. 2014;68:419–30.

[54] Schiewe MC, Gamboa L, Smetona V, Baskevitch K, Anderson RE. Comparative assessment of human blastocyst resiliency to vitrification solution toxicity and osmotic stress associated with re-vitrification (rVTF). J Reprod Biotechnol Fertil. 2016 (accepted).

[55] Fahy GM, Wowk B, Wu J, Paynter S. Improved vitrification solutions based on the predictability of vitrification solution toxicity. Cryobiology. 2004;48:22–35.

[56] Vajta G, Nagy ZP, Cobo A, Conceicao J, Yovich J. Vitrification in assisted reproduction: Myths, mistakes, disbeliefs and confusion. Reprod Biomed Online. 2009;19:1–7.

[57] Schiewe MC. MicroSecure vitrification for oocytes and embryos: Optimum simplicity, security and cost and effectiveness combining FDA-approved products. J Clinical Embryol. 2010;13:33–51.

[58] Schiewe MC. The historic development and incorporation of four assisted reproductive technologies shaping today's IVF industry. J Fertil In Vitro Reprod Med Genet. 2016;4:2–7.

Section 5

Plant Cryopreservation

Chapter 12

Cryopreservation of Orchid Genetic Resources by Desiccation: A Case Study of *Bletilla formosana*

Rung-Yi Wu, Shao-Yu Chang, Ting-Fang Hsieh,
Keng-Chang Chuang, Ie Ting, Yen-Hsu Lai and
Yu-Sen Chang

Additional information is available at the end of the chapter

http://dx.doi.org/10.5772/65302

Abstract

Many native orchid populations declined yearly due to economic development and climate change. This resulted in some wild orchids being threatened. In order to maintain the orchid genetic resources, development of proper methods for the long-term preservation is urgent. Low temperature or dry storage methods for the preservation of orchid genetic resources have been implemented but are not effective in maintaining high viability of certain orchids for long periods. Cryopreservation is one of the most acceptable methods for long-term conservation of plant germplasm. Orchid seeds and pollens are ideal materials for long-term preservation (seed banking) in liquid nitrogen (LN) as the seeds and pollens are minute, enabling the storage of many hundreds of thousands of seeds or pollens in a small vial, and as most species germinate readily, making the technique very economical. This article describes cryopreservation of orchid genetic resources by desiccation and a case study of *Bletilla formosana*. We hope to provide a more practical potential cryopreservation method for future research needs.

Keywords: long-term conservation, *Bletilla formosana*, Desiccation, Dry, Orchid, Seed, Pollen, genetic resources

1. Introduction

Germplasm conservation is mostly applied for breeding purpose. Four methods are usually used in orchid preservation. The first method is more easy to preserve whole plant. It preserves the whole plant in the net-house or greenhouse, most orchid breeders follow this method, but the orchid plants are often lost due to natural disasters, pests, diseases, and physiological disorders during cultivation process. The second method is to preserve orchid cells or tissues by tissue culture. Besides much labor requirements, a lot of problems may occur, such as genetic variation, germplasm pollution, and somatic cell clone variation during the continuous subculture process. The third method, dry storage or low temperature method has been carried out for the preservation of orchid genetic resources [1, 2]. In order to achieve a successful hybridization or a special breeding purpose, orchid breeders must preserve pollens from different flowering parents. Moreover, seeds of some important, economic value, particularly endangered species also need to be preserved. Depending on the equipment, cost, and convenience, orchid breeders often preserve pollens or seeds at 4°C in a refrigerator. However, this method does not get an acceptable result in keeping high viability of certain orchids for long period [3–5]. In addition, dry storage and low temperature methods used in case of many orchid seeds are only for short-term preservation for 1–6 months. Viability of most orchid seeds is significantly reduced after less than 1 year for preservation. Furthermore, the seeds of certain orchid species lose their viability quickly upon desiccation [6, 7]. Therefore, the last method, cryopreservation which is a long-term preservation technique has been researched and developed intensively for the need of orchid genetic resources preservation and the orchid industry. Cryopreservation is one of the most reliable methods for long-term conservation of plant genetic resources, because all metabolic processes and physicochemical changes are arrested at the cryogenic temperature (-196°C) [8, 9]. However, it is usually lethal to expose biological specimens to such low temperatures without any pretreatment because of intracellular freezing [4]. Vitrification and desiccation methods have been often used to preserve seeds by removing water from the cells [9–11] because the water content of plant materials may affect cryopreservation success. Orchid PLB (protocorm like body) conservation by combining encapsulation and dehydration has been suggested [12–14].

Bletilla formosana belongs to genus *Bletilla* in the family Orchidaceae. The species is distributed widely in Taiwan and is renowned for its ornamental value [5, 15, 16]. *B. formosana* is endangered due to the destruction of its habitat and over collection for ornamental use. Therefore, preservation of *B. formosana* is urgent to be proceeded. The purpose of this article is to review the cryopreservation of orchid germplasm, describe a practical method of long-term preservation for *Bletilla formosana* seeds, and to provide potential cryopreservation methods for other orchid species.

2. Cryopreservation

The process of cryopreservation preserves structurally intact living cells and tissues by cooling them to very low temperatures [17]. Cryopreservation is one of the most effective methods for the long-term conservation of plant germplasm at ultra-low temperatures (–196°C) because through it, the vitality of cells is preserved despite the cessation of almost all of their biological activities [8, 9]. During cryopreservation, degradation or somatic mutation phenomenon rarely occurs [4, 8, 9, 18, 19].

The advantages of cryopreservation are as follows:

1. The ability to preserve the vitality and regenerative potential of cells.
2. A requirement for minimal tissue to be effective, resulting in minimal space being used for operation.
3. The prevention of genetic variation and germplasm pollution, and the reduction of somatic cell clone variation rates.
4. The protection against damage from natural disasters, pests, and diseases by using liquid nitrogen (LN) as the storage material.
5. The reduction in labor requirements to accomplish the complicated process of subculture.
6. The possibility of being applied to vegetative propagation plants, nonseed propagation plants, transgenic plants, and gene banks.

3. Cryopreservation of orchid genetic resources: seed and pollen

The main purpose of the long-term preservation of orchid seeds and pollens is to preserve endangered or economically crucial species. Since orchid seeds and pollens are minute, storing many hundreds of thousands of them in a single small vial is possible, making them ideal materials for long-term preservation in LN. Furthermore, most species of orchids germinate readily. Thus, for both of these reasons, cryopreservation is economical and convenient [11]. As reported in [20], maintaining the proper water content (WC) of seeds is critical for successful cryopreservation because excess moisture can result in 'free' water in tissues forming damaging ice crystals during freezing. In most species, exposing biological samples to such low temperatures without any WC pretreatment is typically lethal because of intracellular freezing [4]. Therefore, pretreatment technologies, for example the vitrification and desiccation methods, have been developed [21] to use dehydration for the reduction of the WC of cells and avoid the formation of ice crystals from ultra-low temperature preservation. Prior to ultra-low temperature preservation, suitable pretreatment methods are used to increase the survival rate of the materials to be preserved. Three pretreatments, namely desiccation, vitrification, and encapsulation–dehydration, are typically applied for orchids [13, 21–24]. Pretreatment technologies prior to cryopreservation are still a fancy work to investigate now.

According to the aforementioned reports, three cryopreservation methods are available for orchids.

3.1. Vitrification method

The vitrification technique was introduced by Sakai et al. and is typically used to preserve immature and mature seeds with a higher than average WC for extended periods. Preserved materials are sufficiently dehydrated osmotically by being placed in a high osmolarity vitrification solution (glycerol, dimethyl sulfoxide, and ethylene glycol), which alters their intracellular WC so as to vitrify them through the penetration of cryoprotectants. The chemicals used in this process are toxic. The functions of cryoprotectants are to reduce the amount of freezable water in seed tissue, reduce the freezing temperatures of the intracellular solutes, and inhibit ice nucleation and growth [24–27].

The seeds of some orchids cannot survive, when preserved at cryogenic temperatures even with relatively low WC. For example, mature seeds of *Oncidium flexuosum* (11% WC) and a *Dendrobium* hybrid (13% WC) were unable to germinate or exhibited a low germination percentage after direct immersion in LN [28, 29]. The seeds of these species should be cryopreserved through a vitrification method combined with rapid cooling and rewarming [28–30].

Vitrification has been applied in the cryopreservation of immature and highly hydrated seeds of orchid species from several genera including *Bletilla, Cymbidium, Cyrtopodium, Dendrobium, Doritis, Encyclia, Phaius, Ponerorchis* and *Vanda* [24]. The vitrification method is effective for all types of in vitro material and for not fully dehydrated materials, but the procedure is more complex and high concentrations of cryoprotectants may be toxic to plant tissues [10, 11].

3.2. Desiccation method

The desiccation method is more suitable for mature seeds than for immature seeds. Narrow desiccation refers to when seeds are first dehydrated through slow drying with a controlled desiccation rate under a constant relative humidity (RH) or through rapid drying under a laminar flow chamber or with silica gel or a saturated salt solution of $CaCl_2 \cdot 6H_2O$ (approximately 33% RH) to reduce seed WC before LN preservation. For example, *Bletilla formosana, Caladenia arenicola, Caladenia flava, Caladenia huegelii, Dactylorhiza fuchsii, Dactylorhiza majalis* spp. *praetermissa, Diuris fragrantissima, Diuris laxiflora, Diuris magnifica, Eulophia gonychlia, Eulophia stenophylla, Gymnadenia conopsea, Microtis media* spp. *media, Orchis coriophora* (*Anacamptis coriophora*), *Orchis morio* (*Anacamptis morio*), *Paphiopedilum rothschildianum, Pterostylis recurva, Pterostylis sanguinea, Thelymitra crinita, Thelymitra macrophylla,* and *Dendrobium candidum* (*Dendrobium moniliforme*) are all most successfully preserved through the desiccation method [21, 31–35] (**Table 1**).

A generalized drying method should be included as the storage condition before LN, such as seeds stored in a low RH environment for a period or fresh seeds stored in a refrigerator, then directly immersed in LN. For example, harvested seeds of *Calanthe gorey, Calanthe vestita* var. *rubro oculata* Paxt., *Encyclia cochleata, Angraecum magdalenae, Miltonia flavescens* × *Brassia longissima,* and *Trichopilia tortilis* were stored in a refrigerator for 7 days to 7 years prior to cryopreservation [36]. Harvested seeds of *Dactylorhiza balitica, D. fuchsii, Dactylorhiza incarnata,* and *Dactylorhiza maculata* were stored at 5–6°C for 2–3 weeks, then directly immersed in LN [37]. Harvested seeds of *Disa uniflora, Eulophia Alta, Eulophia streptopetala, Satyrium nepalense* var. *ciliatum, S. nepalense* var. *nepalense,* and *Phalaenopsis equestris* are stored at 2°C without silica gel for 6–12 weeks prior to immersion in LN [34] (**Table 1**).

The critical factor for successful cryopreservation through desiccation is the existence of the proper WC of tissue, which varies among species [38]. However, because the orchid seed is tiny and light, accurately measuring its moisture context is technically difficult [21]. Therefore, numerous orchid seeds are required to determine their moisture content (MC) prior to desiccation [4, 7, 11]. Moreover, seed viability after cryopreservation by desiccation varies among

Species	Preserved material	Pretreatment before LN	Storage duration	Germination%		Reference
				control [z]	cryo.[z]	
Angraecum magdalenae	Mature seed	At refrigerator for 7 days-7 years	1 month	55	80	[36]
Bletilla formosana	MAP[y]mature seed	Direct LN	1 day	77	2	[35]
Bletilla formosana	3MAP mature seed	Air dry dry over silica gel for 24 h	1 day 1 day	77 77	69 69	[35]
Bletilla striata	2, 3, 4MAP immature mature seed	Direct LN	30 min	2 MAP:0.2 3 MAP:25 4 MAP:99	0 12 33	[43]
Bratonia × Miltonia flavescens	Mature seed	At 8°C for 403 days	1 month	100	100	[55]
Caladenia arenicola	Mature seed	Dry over silica gel for 24 h	1 week	31	37	[31]
Caladenia arenicola	Mature seed	Air dry for 1week →20°C	3-24 months	70-83	55-85	[33]
Caladenia flava	Mature seed	Air dry for 1week →20°C	3-24 months	35-80	0-83	[33]
Caladenia huegelii	Mature seed	Air dry for 1week →20°C	3-24 months	63-95	80-97	[33]
Calanthe Gorey	Mature seed	At refrigerator for 7 days-7 years	1 month	59	90	[36]
Calanthe vestita var. rubrooculata Paxt.	Mature seed	At refrigerator for 7 days-7 years	1 month	56	69	[36]
Dactylorhiza balitica	Mature seed	At 5-6°C for 2-3 weeks	1 month	33	39	[37]
Dactylorhiza fuchsii	Mature seed	Dry over a saturated salt solution of $CaCl_2 6H_2O$ (ca. 33% RH)at 16°C	1-12 months	74	51	[6]
Dactylorhiza fuchsii	Mature seed	At 5-6°C for 2-3 weeks	1 month	36	44	[37]
Dactylorhiza incarnata	Mature seed	At 5-6°C for 2-3 weeks	1 month	14	27	[37]
Dactylorhiza maculata	Mature seed	At 5-6°C for 2-3 weeks	1 month	71	42	[37]
Disa uniflora	Mature seed	No wit silica gel at 2°Cfor 6-12 weeks	15 min	61	56	[34]
Diuris fragrantissima	Mature seed	Air dry for 1 week →20°C	3-24 months	<20	<20	[33]
Diuris laxiflora	Mature seed	Air dry for 1 week →20°C	3-24 months	55-85	20-55	[33]
Diuris magnifica	Mature seed	Dry over silica gel for 24 h	1 week	37	44	[31]

Species	Preserved material	Pretreatment before LN	Storage duration	Germination% control [z]	Germination% cryo.[z]	Reference
Encyclia cochleata (*Prosthechea cochleata*)	Mature seed	At refrigerator for 7 days-7 years	1 month	100	100	[36]
Eulophia alta	Mature seed	No wit silica gel at 2°C for 6-12 weeks	15 min	52	47	[34]
Eulophia stenophylla	Mature seed	In wit silica gel at 2°C for 8 months	15 min	68	70	[34]
Eulophia streptopetala	Mature seed	No wit silica gel at 2°C for 6-12 weeks	15 min	63	58	[34]
Gymnadenia conopsea	Mature seed	In wit silica gel at 2°C for 6-12 weeks	15 min	45	46	[34]
Juglans regia (English walnut)	Pollen	1.at -1°Cmin^{-1} by a controlled-rate freezer 2. air dry 24 h 3. air dry 72 h	12 months 12 months 12 months	1. 82-96 2. 60-88 3. 5-54	1. 0-70 2. 48-81 3. 2-40	[23]
Luisia macrantha	Pollen	1. air dry: 0-30 min 2. dry over silica gel:120 min	48 h	1. 65-67 2. ca.67	1. 54 2. 51-52	[44]
Microtis media spp. media	Mature seed	Air dry for 1 week→20°C	3-24 months	30-67	5-63	[33]
Miltonia flavescens×Brassia longissima	Mature seed	At refrigerator for 7 days-7 years	1 month	100	100	[36]
Narcissus St. Keverne	Pollen	AF, AN, PN [x]	3 days 351 days	AF:27 AN:27 PN:27	AF:16(3 days), 0.1(351 days) AN:15(3 days), 13(351 days) PN:16(3 days), 13(351 days)	[56]
Orchis coriophora (*Anacamptis coriophora*)	Mature seed	In wit silica gel at 2°C for 6-12 weeks	15 min	43	43	[34]
Orchis morio (*Anacamptis morio*)	Mature seed	In wit silica gel at 2°C for 6-12 weeks	10 cycle(5 min each)	62	75-82	[34]
Phalaenopsis equestris	Mature seed	No wit silica gel at 2°C for 6-12 weeks	15 min	84	79	[34]
Platanthera bifolia	Mature seed	At 5-6°C for 2-3 weeks	1 month	49	24	[37]
Pterostylis recurva	Mature seed	Air dry for 1week→20°C	3-24 months	70-85	0-90	[33]
Pterostylis sanguinea	Mature seed	Dry over silica gel for 24 h	1 week	49	50	[31]
Pterostylis sanguinea	Mature seed	Air dry for 1 week→20°C	3-24 months	85-90	25-75	[33]

Species	Preserved material	Pretreatment before LN	Storage duration	Germination%		Reference
				control [z]	cryo.[z]	
Satyrium nepalense var. ciliatum	Mature seed	No wit silica gel at 2°C for 6-12 weeks	15 min	58	55	[34]
Satyrium nepalense var. nepalense	Mature seed	No wit silica gel at 2°C for 6-12 weeks	15 min	62	62	[34]
Thelymitra crinita	Mature seed	Dry over silica gel for 24 h	1 week	59	67	[31]
Thelymitra crinita	Mature seed	Air dry for 1week→20°C	3-24 months	25-85	0-80	[33]
Thelymitra macrophylla	Mature seed	Air dry for 1week→20°C	3-24 months	20	3-37	[33]
Trichopilia tortilis	Mature seed	At refrigerator for 7 days-7 years	1 month	83	90	[36]
Vanda pumilla	Mature seed	Not given	15 min	94	97	[34]
Vanda tricolor	Mature seed	Direct LN	1 day	11	1	[22]
Vanda tricolor	Immature seed	Direct LN	1 day	26	10	[22]

[z] Germination of control and cryopreserved seeds, control indicates germination of seeds before cryopreservation.
[y] MAP means months after pollination.
[x] AF: anthers held in a desiccator with calcium chloride at 2°C; AN: anther kept overnight in a desiccator at room temperature→-130°C 1h; PN: anther kept overnight in a desiccator at room temperature→ shake, pollen transferred to straw→-130°C 1h.

Table 1. Cryopreservation of orchid seed or pollen by desiccation prior to LN.

different species. The seeds of some species lose viability after drying to a specific level of WC. Some seeds remain highly viable after being air-dried or exposed to a controlled desiccation rate under a constant RH or to rapid desiccation.

The advantages of the desiccation method are simple, practical, nontoxic, and low cost. Also, the method is more suitable for orchid seed with low WC [24]. Therefore, it can be used for long-term preservation of orchid seed by international plant germplasm centers and private companies.

3.3. Encapsulation-dehydration method

The third cryopreservation method for orchid materials is encapsulation-dehydration. This procedure is based on the technology developed for the production of artificial seeds [39]. Preserved tissues encapsulated in alginate beads, pregrown in a liquid medium with enriched sucrose, partially desiccated in the air current of a laminar airflow cabinet or with silica gel to reduce the WC to a suitable level, and then directly immersed in LN [21].

Among the three aforementioned techniques, the encapsulation-dehydration method is applied only for particular orchid species. For example, pretreatments prior to LN only improved germination of *Cyrtopodium hatschbachii* immature seeds (control/cryopreservation 50/64%), and *Oncidium bifolium* (*Gomes Bifolium*) mature seeds (control/cryopreservation 88/89%) [40, 41] (**Table 1**).

4. Factors affecting cryopreservation

For successful cryopreservation, the required WC of tissue is a critical factor, and varies among species [21, 38]. However, studies have revealed that the survival of orchid seeds in storage is also affected by other factors [7, 42]. For example, the WC of preserved tissue is also affected by the desiccation time and the maturity of the preserved material [43, 44]. An inappropriate desiccation time also reduces the viability of the tissue. Furthermore, the WC of seeds gradually decreases after pollination [43]. The impact of these factors is described as follows.

4.1. Water content

For most orchid species, for example, *Bletilla striata, B. formosana, D. candidum* (*D. moniliforme*), and *Vanda tricolor* [22, 35, 43, 45], preserving tissue immersed directly in LN without any pretreatment results in high mortality rates because of formation of intracellular ice nucleation and growth in the cells [4, 20, 21, 38]. Only a few examples were exceptions, for example, fresh pollens of English walnut (*Juglans regia* L.) described above [23]. As revealed in related studies, various terrestrial and epiphytic species have not been damaged by storage in LN when seed WC was below a critical point [6, 21, 34, 36]. For example, in seeds of *D. candidum*, a high survival rate (approximately 95%) was obtained when the WC of seeds decreased to 8–19% prior to cryopreservation [33]. *E. cochleata* seeds sealed in cryovials with a 24% WC could be cryopreserved without any loss in viability [36]. Fresh mature seeds of *B. formosana*, an endangered Taiwanese orchid, reached a 68% germination rate after being air-dried for 24 h to a 25% WC and cryopreserved through direct immersion in LN [35]. These studies reveal that the WC of orchid seeds required for successful cryopreservation clearly varies among the species.

Pretreatment techniques must be performed before preserving seeds in LN for most orchids. Previous research on plants such as *B. striata, Phaius tancarvilleae*, and *Vanda coerulea* has revealed that preserved tissue must be pretreated prior to cryopreservation to yield high survival rates [4, 11, 18, 43, 46]. Pretreatment is critical and should be applied to most plant tissues before cryopreservation, as stated in the preceding paragraph. Therefore, although cryopreservation has been employed for the long-term preservation of plant tissues for a number of decades, investigation into pretreatment techniques has continued to contribute to improved tissue viability [14, 34, 47].

Some seeds of orchid species have exhibited sensitivity toward extreme desiccation and a reduced viability after short periods of storage under dry conditions (3–5% WC) at high subzero temperatures (e.g., −5°C to −20°C) generally applied in conventional seed banking regimes [4, 24, 36, 45, 48, 49]. The seeds of *P. tancarvilleae* are difficult to preserve with low WC (2–5%) and at 4°C and 25°C [4]. Pollinia of *Luisia macrantha* subjected to 0–30 min of dehydration exhibited maximum germination rates of 65–67% [44]; their germination rates decreased with increasing desiccation time. A high survival rate (approximately 95%) of *Dendrobium candidum* (*Dendrobium moniliforme*) seeds can be achieved by drying seeds to WC of 8–19%; however, seed growth is slower in samples dried to WC less than 12%. This reveals that an optimal WC of 12–19% exists for cryopreservation. Species with such desiccation sensitivity have a different optimal cryopreservation method than the general orchid species.

4.2. Maturity

Successful cryopreservation is closely related to the WC of plant tissue [49]. Furthermore, the WC of seeds is related to seed maturity. For example, the seeds of *B. striata* revealed a steady reduction in WC with an increase in time after pollination: 84, 57, and 33% for 2, 3, and 4 months after pollination (MAP), respectively [43]. The WC of the developing seeds decreased with time after pollination [43]. When these immature seeds were directly immersed in LN without any pretreatment, there was a tendency toward an improved germination rate with an increase in seed age: 0, 11.6, and 32.8% for 2, 3, and 4 MAP, respectively. Those of the three or four MAP samples might have developed an ability to survive the frozen state. Vacuolation has been observed in young embryo cells of orchids [50, 51]. However, it disappears as the orchid tissue approaches maturity. In the aforementioned study of *B. striata*, reduced vacuolation and, consequently, a lower level of bulk water in the embryo cells that accompany seed maturation may have prevented intracellular freezing in immature seeds [43].

The cryopreservation methods also differ depending on the seed maturity level. The desiccation method is more suitable for mature seeds as a cryopreservation pretreatment. The vitrification method or the encapsulation-dehydration method may be used for many immature and highly hydrated orchid seeds. The pretreatment in concentrated cryoprotectant solutions before rapid immersion in LN is critical for post-cryopreservation survival of orchid seeds, for example, *Bletilla*, *Cyrtopodium*, *Doritis*, *Phaius*, *Ponerorchis*, and *Vanda* [4, 10, 11, 22, 24, 41, 43, 52].

4.3. Other factors

In addition to WC and seed maturity, successful cryopreservation is related to the type of orchid species [21]. Studies have shown that the survival of orchid seeds in storage is affected by factors, such as desiccation time [7, 31, 42]. Desiccation time, by directly affecting the level of seed WC, is crucial for the survival of preserved tissue. Inappropriate desiccation time reduces tissue viability. For example, seed germination of *Coelogyne foerstermannii*, *C. rumphii*, and *D. stratiotes* decreased to 1–5% from initial values of 65–96% after drying in equilibrium to a 15% RH and storage at −20°C for 9–12 months. By contrast, *X. undulatum* seeds lost 13% of their germination rate during the same interval [53]. Such varying results to conventional banking conditions have strengthened the requirement for research into the cryostorage behavior of orchid seeds and the seeds of other species [54]. According to Seaton and Hailes [7], silica gel reduces *Cattleya aurantiaca* seed moisture to a level that results in a rapid loss of viability. This contrasts with results that have indicated that desiccating orchid seeds over silica gel for 24 h induces a significant increase in seed germination. The effect of longer seed desiccation durations remains to be determined. For example, the seeds of *P. tancarvilleae* are difficult to preserve with a low WC (2–5%) at 4 and 25°C [4]; however, the germination rates of *B. formosana* dried seeds with a low WC are 68.5% after cryopreservation [35]. Thus, the appropriate WC that orchid seeds require for successful cryopreservation varies among the species.

The values for the pollen germination and WC of the English walnut (*Juglans regia L.*) were the highest before storage. Subsequently, the values decreased along with an increase in storage time at room temperature. However, they did not decrease linearly with time. The ability to germinate was significantly reduced during drying, though some pollen retained some viability until a 3.2% MC [23]. Pollinia of *L. macrantha* subjected to 0–30 min dehydration within a laminar air flow cabinet showed maximum germination of 65–67% in desiccated controls and 54% in LN treated samples. The germination rate decreased with prolonged desiccation time [44].

Long-term preservation of plant genetic resources is not easy. Orchid seeds and pollens are ideal materials suitable for cryopreservation because of the characteristics of tiny volume and low water content. This article describes common pretreatment techniques used in orchid cryopreservation and the factors affecting material viability. Besides, this describes Taiwan's endangered native medicinal and ornamental plants, and the desiccation method applied in a case study of *Bletilla formosana*. The genetic resources of other economic orchids will continue to be tested by the desiccation method in the future. The ultimate object of our study is to provide a more practical potential cryopreservation method and apply in the other economic orchids for sustainable development of orchid industry.

5. Case study for cryopreservation of *Bletilla formosana* seeds (Orchidaceae) by desiccation

B. formosana belongs to genus *Bletilla* in the family Orchidaceae. The species is distributed widely in Taiwan and is known for its medicinal and ornamental value [5, 15, 16]. *B. formosana* is endangered during the destruction of habitat and over collection for ornamental use. Therefore, preservation of *B. formosana* is urgent to be proceeded.

Seeds of *Bletilla* genus had been studied in the pretreatment method prior to cryopreservation. However, the procedure of common vitrification method is more complex and high concentrations of cryoprotectants may be toxic to plant tissues [10, 11]. Therefore, the purpose of this study is to establish a practical method of long-term preservation for *B. formosana* seeds and to provide potential cryopreservation methods for other orchid species.

Mature seeds of *B. formosana* obtained from capsules 3 months after hand-pollination (3 MAP) were collected as the test materials, then dried at room temperature (27 ± 1°C) for 24 h in the laboratory with air conditioning and dried in a sealed container using silica gel for 0–24 h, respectively. Seeds were then stored in LN for 24 h, to investigate the viability rate and the germination rate of seeds after warming.

The viability and germination rates of fresh seeds of *B. formosana* are 89.9 and 76.8%, respectively. The initial water content of fresh seeds is 49.5% but decreasing significantly within 1–4 h by silica gel desiccation and then getting stable to 1.9%, 1 day after desiccation. When fresh seeds placed directly into LN for cryopreservation, the viability and germination rates are suddenly went down to 2.3 and 1.8%, respectively. However, for the fresh seeds pretreated by silica gel desiccation for 24 h or air-dried for 24 h at room temperature before putting in LN, the viability rates dramatically increased to 86.8 and 84.9%, respectively, and germination rates are up to 68.5 and 68.6%, respectively (**Figure 1**). These data show that the seed viability of *B. formosana* after cryopreservation is affected by the water content of storage seeds. When the water content of orchid seeds decreases to 24.8% by 24 h air drying or to 21.9–31.2% by 1–2 h silica gel drying, high viability and germination rates still remain after long-term

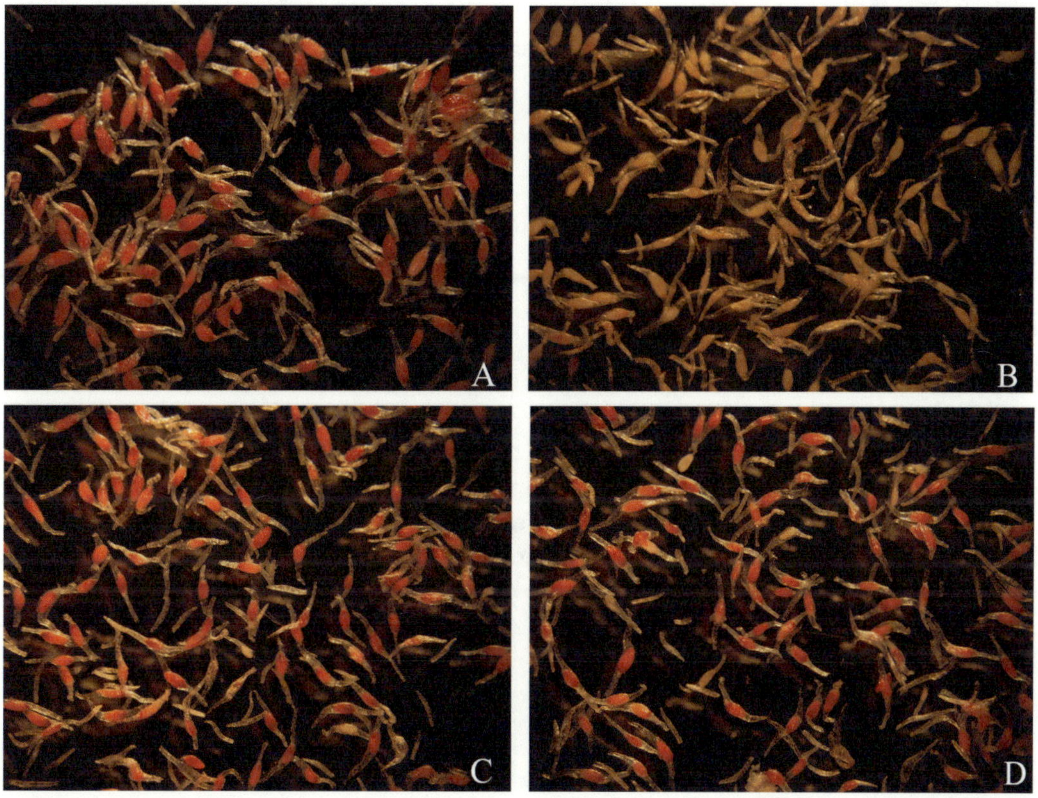

Figure 1. Effect of desiccation on viability of *Bletilla formosana* seeds after cryopreservation by TTC staining method. (A) Harvested fresh seeds; (B) fresh seeds plunged into liquid nitrogen (LN, -196°C); (C) fresh seeds dried for 24 h with silica gel prior to LN; (D) fresh seeds dried for 24 h with air-drying in laboratory conditions prior to LN. Bar =1 mm [49].

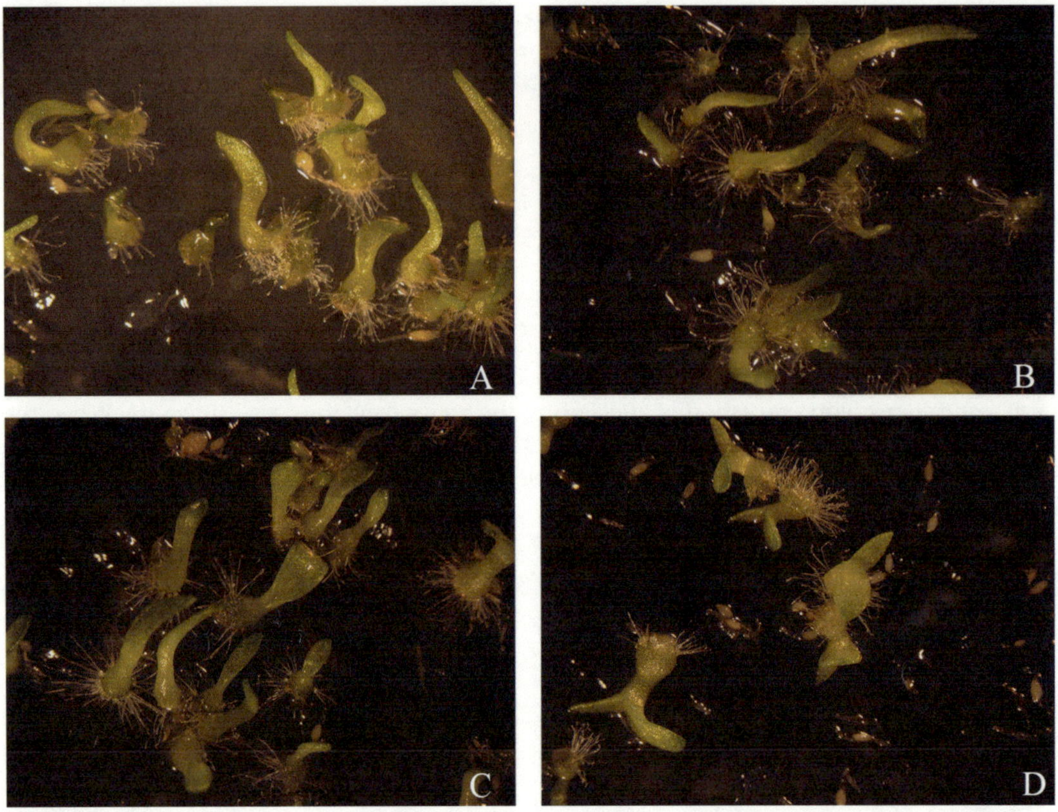

Figure 2. Seedling growth of *Bletilla formosana* for 6 wk after sowing on MS medium. (A) Fresh seeds; (B) fresh seeds were dried for 24 h with air-drying in laboratory conditions prior to LN; (C) fresh seeds were dried for 24 h with silica gel prior to LN; (D) fresh seeds were dried for 24 h with silica gel, vitrification (0.06 M sucrose solution+LS+PVS2 1 h) prior to LN. Bar =1 mm [49].

cryopreservation. Therefore, those desiccation methods are recommended for long term storage of *B. formosana*. Fresh seeds of orchid pretreated in sucrose solution or vitrification before LN treatment are unsuitable for cryopreservation in our previous study [35]. Seedlings derived from the seeds desiccated by these two methods and then preserved in LN grew well 6 weeks after seed sowing (**Figure 2**). In addition, *B. formosana* seeds with 1.9–24.8% water content were found to be suitable for cryopreservation [35].

Acknowledgments

The authors would like to thank Professor Fure-Chyi Chen (National Pingtung University of Science and Technology) and Associate Professor Chen Chang (National Chung Hsing University) for assistance and guidance on the experiment.

http://dx.doi.org/10.5772/65302

Author details

Rung-Yi Wu[1], Shao-Yu Chang[1], Ting-Fang Hsieh[1], Keng-Chang Chuang[1], Ie Ting[1], Yen-Hsu Lai[2] and Yu-Sen Chang[3]*

*Address all correspondence to: yschang@ntu.edu.tw

1 Floriculture Research Center, Taiwan Agricultural Research Institute, COA, Yunlin, Taiwan

2 Department of Horticulture, National Tseng-Wen Senior Agricultural & Industrial Vocational School, Tainan, Taiwan

3 Department of Horticulture and Landscape Architecture, National Taiwan University, Taipei, Taiwan

References

[1] Arditti J, Emst R. Micropropagation of Orchid. Wiley. 1993. New York, U.S.A.

[2] Sivasubramaniam S, Loh CS, Lee YK, Hew CS. Low temperature storage of *Dendrobium* Multico White callus tissue. Malayan Orchid Rev. 1987. 21:22–26.

[3] Chang C, Sung PG, Chang CH, Chen YC, Lin YH. Seed development and storage of *Bletilla formosana* (Hayata) Schltr. Seed Nursery. 2006. 8:29–38.

[4] Hirano T, Godo T, Miyoshi K, Ishikawa K, Ishikawa M, Mii M. Cryopreservation and low-temperature storage of seeds of *Phaius tankervilleae*. Plant Biotechnol. Rep. 2009. 3:103–109.

[5] Lin YJ, Chen CC, Yeh FT, Chiu NY, Tsay HS. Tissue culture of *Bletilla formosana* I. The influence of seed maturity and pretreatment on seed germination and seedling development. J. Agric. Res. China. 1994. 43:40—50.

[6] Pritchard HW, Poynter ALC, Seaton PT. Interspecific variation in orchid seed longevity in relation to ultra-dry storage and cryopreservation. Lindleyana. 1999. 14:92–101.

[7] Seaton PT, Hailes NSJ. Effect of temperature and moisture content on the viability of *Cattleya aurantiaca* seed, in: Pritchard, H.W. (Ed.), Modern Methods in Orchid Conservation: The Role of Physiology, Ecology and Management. Cambridge University, Cambridge, U.K., 1989. pp. 17—29.

[8] Engelmann F. Importance of crop for conservation of plant genetic resources, in: Engelmann F, Takagi H. (Eds.), Cryopreservation of Tropical Plant Germplasm. JIRCAS, Tsukuba, Japan, 2000. pp. 8—20.

[9] Gonzalez-Arnao MT, Panta A, Roca WM, Escobar RH, Engelmann F. Development and large scale application of cryopreservation techniques for shoot and somatic embryo cultures of tropical. Plant Cell Tissue Organ Cult. 2008. 92:1–13.

[10] Ishikawa K, Harata K, Mii M, Sakai A, Yoshimatsu K, Shimomura K. Cryopreservation of zygotic embryos of a Japanese terrestrial orchid (*Bletilla striata*) by vitrification. Plant Cell Rep. 1997. 16:754–757.

[11] Thammasiri K. Cryopreservation of seeds of a Thai orchid (*Doritis pulcherrima* Lindl.) by vitrification. Cryo Letters. 2000. 21:237–244.

[12] Antony JJJ, Sinniah UR, Keng CL, Pobathy R, Khoddamzadeh AA, Subramaniam S. Selected potential encapsulation-dehydration parameter on *Dendrobium* Bobby Messina protocorm-like bodies using TTC analysis. Aust. J. Crop Sci. 2011. 5:1817–1822.

[13] Khoddamzadeh AA, Sinniah UR, Lynch P, Kadir MA, Kadzimin SB, Mahmood M. Cryopreservation of protocorm-like bodies (PLBs) of *Phalaenopsis bellina* (Rchb.f.) Christenson by encapsulation-dehydration. Plant Cell Tissue Organ Cult. 2011. 107:471–481.

[14] Subramaniam S, Sinniah UR, Khoddamzadeh AA, Periasamy S, James JJ. Fundamental concept of cryopreservation using *Dendrobium* sonia-17 protocorm-like bodies by encapsulation-dehydration technique. Afr. J. Biotechnol. 2011. 10:3902–3907.

[15] Yeh FT, Lin YC, Chen CC, Chiu NY, Tsay HS. Tissue culture of *Bletilla formosana* (Hayata) Schlechter II. The influence of medium components on immature seed germination and seedling development. Chin. Agron. J. 1995. 5:353–363.

[16] Ying SS. Coloured illustrated orchid flora of Taiwan Vol. I, first ed. S.–S. Ying Publisher. Taipei, Taiwan. 1996.

[17] Pegg DE. Principles of cryopreservation. Methods Mol Biol. 2007. 368:39–57.

[18] Matsumoto T, Mochida K, Itamura H, Sakai A. Cryopreservation of persimmon (*Diospyros kaki* Thunb.) by vitrifiation of dormant shoot tips. Plant Cell Rep. 2001. 20:398–402.

[19] Na HY, Kondo K. Cryopreservation of tissue-cultured shoot primordia from shoot apices of cultured protocorms in *Vanda pumila* following ABA preculture and desiccation. Plant Sci. 1996. 118:195–201.

[20] Stanwood PC, Bass LN. Seed germplasm preservation using liquid nitrogen. Seed Sci. Technol. 1981. 9:423—437.

[21] Hirano T, Ishikawa K, Mii M. Advances on orchid cryopreservation. p.410-414. in: Floriculture Ornamental and Plant Biotechnology: Advances and Topical Issues. Vol. 2. (Teiceira da Silva, J.A., ed.) Global Science Book, 1st ed. Ikenobe, London, U.K. 2006.

[22] Jitsopakul N, Thammasiri K, Yukawa T, K Ishikawa. Effect of cryopreservation on seed germination and protocorm development of *Vanda tricolor*. Sci. Asia. 2012. 38:244–249.

[23] Luza JG, Polito VS. Cryopreservation of English walnut (*Juglans regia* L.) pollen. Euphytica. 1988. 37:141–148.

[24] Popova E, Kim HH, Saxena PK, Engelmann F. Frozen beauty: The cryobiotechnology of orchid diversity. Biotechnol. Adv. 2016. 34:380–403.

[25] Finkle BJ, Zavala ME, Ulrich JM. Cryoprotective compounds in the viable freezing of plant tissues, in: Kartha, K.K. (Ed.), Cryopreservation of Plant Cells and Organs. CRC Press Inc., Boca Raton, Florida, 1985. pp. 75–113.

[26] Fuller BJ. Cryoprotectants: the essential antifreezes to protect life in the frozen state. Cryo Letters. 2004. 25:375–388.

[27] Volk GM, Walters C. Plant vitrification solution 2 lowers water content and alters freezing behavior in shoot tips during cryoprotection. Cryobiology. 2006. 52:48–61.

[28] Galdiano Jr, RF, Lemos EGM, Faria RT, Vendrame WA. Cryopreservation of Dendrobium hybrid seeds and protocorms as affected by phloroglucinol and Supercool X 1000. Sci. Hortic. 2012. 148:154–160.

[29] Galdiano Jr, Lemos RF, EGM, Vendrame, WA. 2013. Cryopreservation, early seedling development, and genetic stability of *Oncidium flexuosum* Sims. Plant Cell Tissue Organ Cult. 114:139–148.

[30] Galdiano Jr, RF, de Macedo Lemos, EG, de Faria, RT, Vendrame, W.A., 2014. Seedling development and evaluation of genetic stability of cryopreserved Dendrobium hybrid mature seeds. Appl. Biochem. Biotechnol. 172:2521–2529.

[31] Batty AL, Dixon KW, Brundrett M, Sivasithamparam K. Long-term storage of mycorrhizal fungi and seed as a tool for the conservation of endangered Western Australian terrestrial orchids. Aust. J. Bot. 2001. 49:619–628.

[32] Pritchard HW, Poynter ALC, Seaton PT. Interspecific variation in orchid seed longevity in relation to ultra-dry storage and cryopreservation. Lindleyana. 1999. 14:92–101.

[33] Hay FR, Merritt DJ, Soanes JA, Dixon KW. Comparative longevity of Australian orchid (Orchidaceae) seeds under experimental and low temperature storage conditions. Bot. J. Linn. Soc. 2010. 164:26–41.

[34] Pritchard HW. Liquid nitrogen preservation of terrestrial and epiphytic orchid seeds. Cryo Letters. 1984. 5:295–300.

[35] Wu RY, Chang SY, Hsieh TF, Chang YS. Cryopreservation of *Bletilla formosana* seeds (Orchidaceae) by desiccation. Sci. Hortic. 2013. 157:108–112.

[36] Nikishina TV, Popov AS, Kolomeitseva GL, Golovkin BN. Effect of cryopreservation on seed germination of rare tropical orchids. Russ. J. Plant Physiol. 2001. 48:810–815.

[37] Nikishina TV, Popova EV, Vakhrameeva MG, Varlygina TI, Kolomeitseva GL, Burov AV. 2007. Cryopreservation of seeds and protocorms of rare temperate orchids. Russ. J. Plant Physiol. 2007. 54:121–127.

[38] Sakai A. Development of cryopreservation techniques, in: Engellmann F, Takagi H (eds.), Cryopreservation of Tropical Plant Germplasm. IPGRI, Italy. 2000. pp. 1–7.

[39] Fabre J, Dereuddre J. Encapsulation-dehydration: a new approach to cryopreservation of Solanum shoot tips. Cryo Letters. 1990. 11:413–426.

[40] Flachsland E, Terada G, Scocchi A, Rey H, Mroginski L, Engelmann F. Cryopreservation of seeds and in vitro-cultured protocorms of *Oncidium bifolium sims*.(Orchidaceae) by encapsulation-dehydration. ?ryo Letters. 2006. 27:235–242.

[41] Surenciski MR, Flachsland EA, Terada G, Mroginski LA, Rey HY. Cryopreservation of *Cyrtopodium hatschbachii* Pabst (Orchidaceae) immature seeds by encapsulation-dehydration. Biocell. 2012. 36:31–36.

[42] Miyoshi K, Mii M. Stimulatory effects of sodium and calcium hypochlorite, pre-chilling and cytokinins on the germination of *Cypripedium macranthos* seed in vitro. Physiol. Plant. 1998.102:481–486.

[43] Hirano T, Godo T, Mii M, Ishikawa K. Cryopreservation of immature seeds of *Bletilla striata* by vitrification. Plant Cell Rep. 2005. 23:534–539.

[44] Ajeeshkumar S, Decruse SW. Fertilizing ability of cryopreserved pollinia of *Luisia macrantha*, an endemic orchid of Western Ghats. Cryo Letters. 2013. 34:20–29.

[45] Wang JH, Ge JG, Liu F, Bian HW, Huang CN. Cryopreservation of seed and protocorms of *Dendrobium candidum*. Cryo Letters. 1998. 19:123–128.

[46] Thammasiri K, Soamkul L. Cryopreservation of *Vanda coerulea* Griff. ex Lindl. seeds by vitrification. Sci. Asia. 2007. 33:223–227.

[47] Kendall EJ, Qureshi JA, Kartha KK, Leung N, Chevrier N, Caswell K, Chen THH. Regeneration of freezing-tolerant spring wheat (*Triticum aestivum* L.) plants from cryoselected callus. Plant Physiol. 1990. 94:1756–1762.

[48] Cherevchenko TM, Kushnir GP. Orkhideiv Kulture (Orchids in Culture). Naukova Dumka, Kiev. 1986 (in Russian).

[49] Koopowitz H, Thornhill A. Gene banking and orchid seeds. Am. Orchid Soc. Bull. 1994. 63:1383–1386.

[50] Manning JC, Van Staden J. The development and mobilization of seed reserves in some African orchids. Aust. J. Bot. 1987. 35:343–353.

[51] Yeung EC, Law SK. Embryology of *Calypso bulbosa*. II. Embryo development. Can. J. Bot. 1992. 70:461–468.

[52] Hirano T, Ishikawa K, Mii M. Cryopreservation of immature seeds of *Ponerorchis graminifolia* var. *suzukiana* by vitrification. Cryo Letters. 2005. 26:139–146.

[53] Seaton P, Kendon JP, Pritchard HW, Puspitaningtyas DM, Marks TR. Orchid conservation: the next ten years. Lankestriana. 2013. 13: 93–101.

[54] Li DZ, Pritchard HW. The science and economics of ex situ plant conservation. Trends Plant Sci. 2009. 14:614–621.

[55] Popova EV, Nikishina TV, Kolomeitseva GL, Popov AS. The effect of seed cryopreservation on the development of protocorms by the hybrid orchid Bratonia. Russ. J. Plant Physiol. 2003. 50:672–677.

[56] Bows SA. Long-term storage of Narcissus anthers and pollen in liquid nitrogen. Euphytica. 1990 48:275–278.